Termination and Connection of Conductors

Malcolm Doughton and John Hooper

CENGAGE
Learning®

Australia • Brazil • Japan • Korea • Mexico • Singapore • Spain • United Kingdom • United States

CENGAGE
Learning·

Termination and Connection of Conductors
Malcolm Doughton and John Hooper

Publishing Director: Linden Harris

Commissioning Editor: Lucy Mills

Editorial Assistant: Claire Napoli

Project Editor: Alison Cooke

Production Controller: Eyvett Davis

Marketing Manager: Lauren Mottram

Typesetter: S4Carlisle Publishing Services

Cover design: HCT Creative

Text design: Design Deluxe

For product information and technology assistance, contact **emea.info@cengage.com**.

For permission to use material from this text or product, and for permission queries, email **emea.permissions@cengage.com**.

British Library Cataloguing-in-Publication Data
A catalogue record for this book is available from the British Library.

ISBN: 978-1-4080-3994-6

Cengage Learning EMEA
Cheriton House, North Way, Andover, Hampshire, SP10 5BE, United Kingdom

Cengage Learning products are represented in Canada by Nelson Education Ltd.

For your lifelong learning solutions, visit **www.cengage.co.uk**

Purchase your next print book, e-book or e-chapter at **www.cengagebrain.com**

Printed in Malta by Melita Press
1 2 3 4 5 6 7 8 9 10 – 15 14 13

Dedication

This series of study books is dedicated to the memory of Ted Stocks whose original concept, and his publication of the first open learning material specifically for electrical installation courses, forms the basis for these publications. His contribution to training has been an inspiration and formed a solid base for many electricians practising their craft today.

The Electrical Installation Series

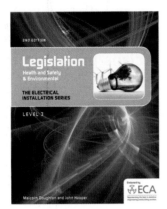

Legislation: Health and Safety & Environmental

Organizing and Managing the Work Environment

Principles of Design Installation and Maintenance

Installing Wiring Systems

Planning and Selection for Electrical Systems

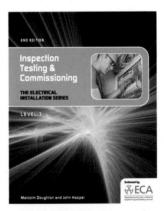

Inspection Testing and Commissioning

Fault Finding and Diagnosis

Maintaining Electrotechnical Systems

Contents

About the authors

Malcolm Doughton

Malcolm Doughton, I.Eng, MIET, LCG, has experience in all aspects of electrical contracting and has provided training to heavy current electrical engineering to HNC level. He currently provides training on all aspects of electrical installations, inspection, testing, and certification, health and safety, PAT and solar photovoltaic installations. In addition, Malcolm provides numerous technical articles and is currently managing director of an electrical consultancy and training company.

John Hooper

John Hooper spent many years teaching a diverse range of electrical and electronic subjects from craft level up to foundation degree level. Subjects taught include: Electrical Technology, Engineering Maths, Instrumentation, P.L.C.s, Digital, Power and Microelectronic Systems. John has also taught various electrical engineering subjects at both Toyota and JCB. Prior to lecturing in further and higher education he had a varied career in both electrical engineering and electrical installations.

Acknowledgements

The authors and publisher would like to thank Chris Cox and Charles Duncan for their considerable contribution in bringing this series of study books to publication. We extend our grateful thanks for their unstinting patience and support throughout this process.

The authors and publisher would also like to thank the following for providing pictures for the book:

AMI
Cable Monkey
Cable Terminology Ltd
Cable Tooling
Cooper Bussman
Draper Tools Ltd
Electrium Sales Ltd.
Electronics Weekly
Fastenright Ltd
Greenlee Textron Inc
Hager Ltd
Kewtech
Labgear

Martindale
MK Electric
On Demand Supplies
Prysmian
Reece Safety
Ripley Tools
Screw Fix
TLC Direct
Top Cable Accessories
Trend Control Systems Ltd
Triton Showers
www.ultimatehandyman.co.uk

Every effort has been made to contact the copyright holders.

This book is endorsed by:

Representing the best in electrical engineering and building services

Founded in 1901, the Electrical Contractors' Association (ECA) is the UK's leading trade association representing the interests of contractors who design, install, inspect, test and maintain electrical and electronic equipment and services.

www.eca.co.uk

Study guide

This study book has been written and compiled to help you gain the maximum benefit from the material contained in it. You will find prompts for various activities all the way through the study book. These are designed to help you ensure you have understood the subject and keep you involved with the material.

Where you see 'Sid' as you work through the study book, he is there to help you, and the activity 'Sid' is undertaking will indicate what it is you are expected to do next.

Task

Familiarize yourself with the requirements of BS 7671, Section 526 Electrical Connections before you continue with this chapter.

Task A 'Task' is an activity that may take you away from the book to do further research either from other material or to complete a practical task. For these tasks you are given the opportunity to ask colleagues at work or your tutor at college questions about practical aspects of the subject. There are also tasks where you may be required to use manufacturers' catalogues to look up your answer. These are all important and will help your understanding of the subject.

Try this

State all the environmental hazards likely to be present when cables are to be terminated in the following locations:

1   A motor isolator in a cement processing works

_____

2   A luminaire in a carwash

_____

3   A street sign at the seafront

_____

Try this A 'Try this' is an opportunity for you to complete an exercise based on what you have just read, or to complete a mathematical problem based on one that has been shown as an example.

Remember

It is essential to terminate this type of cable in a dry atmosphere as the insulation will absorb moisture.

Remember A 'Remember' box highlights key information or helpful hints.

RECAP & SELF ASSESSMENT

Circle the correct answers.

1 The statutory document which identifies a requirement for the quality of terminations is:

   a. BS 7671

   b. The Building Regulations

   c. Electricity at Work Regulations

   d. Health and Safety at Work (etc) Act

2 The Figure 5 in BS 7671 Regulation 526.1 indicates that the regulation can be found in:

   a. Part 5

   b. Section 5

   c. Chapter 5

   d. Regulation 5

Recap & Self Assessment At the beginning of all the chapters, except the first, you will be asked questions to recap what you learned in the previous chapter. At the end of each chapter you will find multichoice questions to test your knowledge of the chapter you have just completed.

Note

Health and Safety Guidance Note GS 38, Electrical test equipment for use by electricians is available as a free download from www.hse. gov.uk.

Note 'Notes' provide you with useful information and points of reference for further information and material.

This study book has been divided into Parts, each of which may be suitable as one lesson in the classroom situation. If you are using the study book for self tuition then try to limit yourself to between 1 hour and 2 hours before you take a break. Try to end each lesson or self study session on a Task, Try this or the Self Assessment Questions.

When you resume your study go over this same piece of work before you start a new topic.

Where answers have to be calculated you will find the answers to the questions at the back of this book but before you look at them check that you have read and understood the question and written the answer you intended to. All of your working out should be shown.

At the back of the book you will also find a glossary of terms which have been used in the book.

A 'progress check' at the end of Chapter 3, and an 'end test' covering all the material in this book, are included so that you can assess your progress.

There may be occasions where topics are repeated in more than one book. This is required by the scheme as each unit must stand alone and can be undertaken in any order. It can be particularly noticeable in health and safety related topics. Where this occurs read the material through and ensure that you know and understand it and attempt any questions contained in the relevant section.

You may need to have available for reference current copies of legislation and guidance material mentioned in this book. Read the appropriate sections of these documents and remember to be on the lookout for any amendments or updates to them.

Your safety is of paramount importance. You are expected to adhere at all times to current regulations, recommendations and guidelines for health and safety.

Unit six

Termination and connection of conductors

Material contained in this studybook covers the knowledge requirement for C&G Unit No. 2357-306 (ELTK 05), and the EAL unit QELTK3/0045.

Unit five considers the practices and procedures for the termination and connection of conductors, cables and flexible cords in electrical wiring systems and equipment. It considers the procedures and applications of different methods of terminating and connecting cables and flexible cords. It also covers the regulatory requirements which apply to the terminations and connections.

You could find it useful to look in a library or online for copies of the legislation and guidance material mentioned in this unit. Read the appropriate sections and remember to be on the lookout for any amendments or updates to them. You will also need to have access to manufacturers' catalogues for wiring systems, tools and fixings.

Before you undertake this unit read through the study guide on pages viii-ix. If you follow the guide it will enable you to gain the maximum benefit from the material contained in this unit.

1

Safe isolation of conductors

Carrying out and ensuring safe isolation is an essential part of an electrician's work. For this reason the requirements for safe isolation appear in a number of the units of the national occupational standard. If you have already covered the subject in another study book in this series then check that you remember and understand what you have already learned. You will need to attempt all the tasks and questions as they will be different for each study book.

LEARNING OBJECTIVES

On completion of this chapter you should be able to:

● Specify the correct procedure for completing safe isolation.

● State the implications of not carrying out safe isolations to:

  – Self

  – Other personnel

  – Customers/clients

  – Public

  – Building systems (loss of supply).

● State the implications of carrying out safe isolations to:

- Other personnel

- Customers/clients

- Public

- Building systems (loss of supply).

Part 1 The requirement for safe isolation

This chapter considers the requirement for safe isolation of electrical circuits and installations to enable electrical work to be carried out safely.

Whilst working through this chapter you will need to refer to Health and Safety Guidance Note GS 38, Electrical test equipment for use by electricians.

> **Note**
> Health and Safety Guidance Note GS 38, Electrical test equipment for use by electricians is available as a free download from **www.hse.gov.uk**.

During the course of our electrical installation work there are many occasions where we will be required to work on a circuit which has been placed in service. This is often to allow us to make alterations or additions to a circuit or installation. When undertaking electrical work it is important to ensure that the part of the installation we are going to work on is safely isolated from the supply. Safe isolation does not simply mean making sure that the supply is switched off; it also includes making sure that it is not inadvertently re-energized.

Figure 1.1 *Safe isolation is required*

There are a number of reasons why safe isolation must be carried out and this action will have implications for building systems, other people and ourselves. Similarly, failure to carry out safe isolation will also have its implications and perhaps it would be as well to consider these first.

Failure to safely isolate

The most obvious implication of the failure to safely isolate when working on electrical installations and equipment is the risk of electric shock to ourselves and others.

The effect of electric current on the bodies of humans and animals is well recorded. The values quoted here are generic and so should be taken as general guidance. A current across the chest of a person in the region of 0.05A (50 mA) and above is enough to produce ventricular fibrillation of the heart which may result in death.

As the average human body resistance is considered to be in the region of 1KΩ (1000Ω) with a voltage of 230V then the current would be in the region of 0.23A (230mA). That is over 4.5 times the level needed to cause ventricular fibrillation.

When working on the electrical installation of equipment we often have to expose live parts, which if not safely isolated, pose a serious risk of electric shock. Other people and livestock within the vicinity of our work will also be able to access these live parts and may not have sufficient knowledge and understanding to avoid the dangers involved.

Where our work is carried out in public areas this risk is further heightened as the installation and equipment may be accessed by anyone: adults, children and animals. The failure to safely isolate presents a very real danger.

Figure 1.2 *Safe isolation is important in busy public areas like shopping centres*

Task

1   Using the information available on the internet look up the definition and details of ventricular fibrillation.

_____

2   State

   a   which area of the heart is affected when ventricular fibrillation occurs

   _____

   b   the next stage if assistance is not given

   _____

   c   the likely outcome if no treatment is provided

   _____

Electric shock also carries the danger of electrical burns which occur at the entry and exit points of the contact and within the body along the path taken by the current. These burns can be severe and whilst the casualty may survive the electric shock the damage, some of which may be irreparable, can be considerable.

Figure 1.3 *An electric shock can cause serious injury*

Whilst an electric shock may not be severe or fatal there is a risk of further injury as a result of a shock. Injuries which may occur include falls from steps and ladders, injury from machinery and vehicles, all of which are caused by the involuntary reaction when a person receives an electric shock.

Figure 1.4 *Failure to safely isolate can lead to serious problems*

To ensure these dangers are removed safe isolation of the installation, circuit or equipment is essential.

Failure to isolate can also affect the building and structure. Failure to isolate introduces a risk of arcing where live parts are exposed. This may occur between live parts at different potentials (line to neutral and between line conductors) and between live conductors and earth.

When an arc is produced electrical energy is converted into heat energy and the level of discharge energy results in a molten conductor being present in the arc. This presents a real risk of fire, and isolation of the circuit(s) by operation of the protective device will not extinguish a fire started in this way.

Note

A single ELV torch battery can produce sufficient energy in a spark to cause an explosion or fire in a flammable or explosive atmosphere such as areas where flammable vapours, dust or gases are present.

In the case of a fault, including one which does not result in an electric shock or fire, the supply should be disconnected automatically. In this case the circuit or installation may be disconnected unexpectedly. The building systems may be switched off resulting in loss of data, failure of heating or ventilation, lighting and power.

In a severe case we may also lose the building and life protection systems such as fire alarm systems, sprinklers, smoke vents, firefighters' lifts and the like. It may also result in the loss of lighting and ventilation to internal areas that have no natural light or ventilation.

This may result in considerable expense to the client and damage to the electrical equipment and buildings. The loss of lighting can cause other dangers to persons within the building resulting in trips, falls and injuries from machinery and equipment.

Figure 1.5 *Lack of lighting can be hazardous in dark corridors and stairwells*

Safe isolation

Carrying out safe isolation is essential to safeguard against the dangers we have identified above but there are a number of implications which must be considered.

However, before we isolate we need to consider the effect this will have on persons, the building and the equipment and services. We need to determine the extent of the installation which needs to be isolated to carry out our work safely. There are certain actions which will need the complete installation to be isolated and others where the isolation of one or more circuits may be all that is necessary.

Figure 1.6 *Work could come to a standstill!*

The isolation of the complete installation has serious implications for the users of the installation as electrical equipment and lighting will not be available for use. This means that the timing and duration of the work must be carefully considered and discussed with the user to minimize disruption.

This will also affect our operation as there will be no supply available for lighting or power tools. So we will need to consider task lighting and power for our work and it may be necessary to arrange a temporary supply for the client's equipment.

Figure 1.7 *Temporary installations can be supplied through transformers from the supply company's mains or from the generators on site*

Where any safety services or alarms may be affected we need to consider the consequences of isolation. For example, burglar alarms may be linked to the police or a security company and the loss of supply may result in a response visit. If this is a false alarm the client may be charged for the wasted visit. Similarly, a fire alarm isolated from the supply may cause an alarm to be triggered which could cause the fire service to respond.

In addition there are a number of other building services which may be affected including time locks, door access systems, bar code readers and tills, security cameras and public address systems.

Isolation of individual circuits may also cause inconvenience to the client and the requirements

Figure 1.8 *Security cameras*

will need to be discussed to ensure that any disruption is kept to a minimum. In any event we must always obtain permission before we isolate.

In certain instances it may not be possible to carry out some or all of the work during normal working hours and arrangements will need to be made with the client to ensure that access is permitted. This may involve the client in the provision of staff to attend, for security or access purposes, during the period when the work is carried out.

Task

1   You are carrying out some research in your local or college library when the supply is suddenly isolated. List the equipment and personnel which will be affected by this supply isolation.

2   Using GS 38 for reference, list the five additional dangers which may be present in the working environment that must be considered before any testing is carried out:

a _____

b _____

c _____

d _____

e _____

Part 2 Safe isolation procedures

The key to safety is to follow the correct procedure(s) throughout the isolation process. We will first consider the process for safe isolation of a single circuit so that we can work on it safely.

Let's consider a situation where we are to carry out an alteration and addition to a lighting circuit in a shop. Before we can begin to work on the alteration and addition we must safely isolate the circuit from the supply. This circuit is protected by a BS EN 60898 type C circuit breaker in the shop's distribution board located at the origin of the installation.

Figure 1.9 *Distribution board*

Before we can begin we must first obtain permission to isolate the circuit. This must be obtained from the person responsible for the electrical installation (the duty holder), not just any employee. As we are going to be isolating the supply the duty holder must ensure that the safety of persons and the operation of the business is not going to be compromised. To do this, the area to be affected and the duration of the isolation, should be explained to the duty holder to help in making the decision.

Once we have been given permission to isolate the circuit we must correctly identify the particular circuit within the distribution board. Where there are a number of lighting circuits it is important that we isolate the correct circuit. Providing the distribution board has been correctly labelled and the appropriate circuit charts are available this should be relatively straightforward.

There are proprietary test instruments which allow the identification of a circuit before it is isolated. These rely on being connected to the circuit when they then transmit a signal through the circuit conductors. A second unit is used to sense the signal at the distribution board to identify the fuse or circuit breaker. The sensitivity of the unit can be adjusted to give a clear and reliable indication of the circuit to be isolated.

This works well with circuits which include a socket outlet as the sender unit can be readily plugged into the circuit. However, where the circuit does not include a suitable socket a connection would need to be made to live parts. This introduces a higher risk to the operative as, due to the purpose of the device, the circuit will not be isolated. In these circumstances the operation would require two persons, one to operate the sensing unit and the other to make the connection to the circuit using test probes complying with GS 38.

Extreme care is required when accessing the live terminals to make this connection and it should only be carried out by skilled persons using suitable equipment.

Having correctly identified the circuit the circuit breaker is switched off, isolating the circuit from the supply.

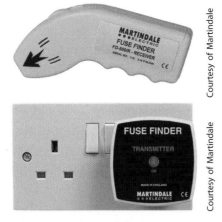

Courtesy of Martindale

Courtesy of Martindale

Courtesy of Martindale

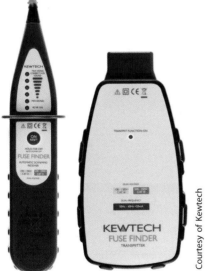

Courtesy of Kewtech

Figure 1.10 *Typical circuit identification instruments*

An appropriate locking off device is then fitted to the circuit breaker to prevent unintentional re-energizing of the circuit. There are a number of proprietary devices available for this task and they all perform the same function: preventing the operation of the circuit breaker. In most cases a separate padlock is inserted through

the locking off device to prevent the unauthorized removal. This secures the circuit breaker and a warning label should also be fitted to advise that the circuit should not be energized and that someone is working on the circuit. Typical lock off devices and labels are shown in Figures 1.11 to 1.13.

Figure 1.11 *Heavy duty lockout tags*

Ideal Industries

Figure 1.12 *Typical lock off kit*

Reece Safety

Figure 1.13 *Locking off devices in place*

Having safely isolated and locked off the circuit we must now confirm that the circuit is indeed isolated from the supply. To do this we will need an approved voltage indicator together with a proving unit, as shown in Figure 1.14. The term 'approved voltage indicator' refers to a voltage indicator which meets all the requirements of Health and Safety Guidance GS 38.

Figure 1.14 *Typical proving unit and approved voltage indicator*

Courtesy of Kewtech

In this instance we are going to remove the cover from a ceiling rose on the circuit at the point where we are going to start the alteration and confirm that it is actually isolated. The first step, having removed the cover, is to confirm that the approved voltage indicator is functioning using the proving unit. The output produced by the proving unit should cause all the LED indicators to light showing that they are all functioning correctly.

As this is a single phase lighting circuit we will need to confirm isolation by testing between:

- All line conductors and neutral (loop line terminal and switch line terminal)
- All line conductors and earth (loop line terminal and switch line terminal)
- Neutral and earth.

And there should be no voltage present at any of these connections.

Task

Using manufacturers' websites and/or catalogues identify five different makes of GS38 compliant 2-pole voltage detection instruments and the rated ac/dc voltage range for each one.

1 _____

2 _____

3 _____

4 _____

5 _____

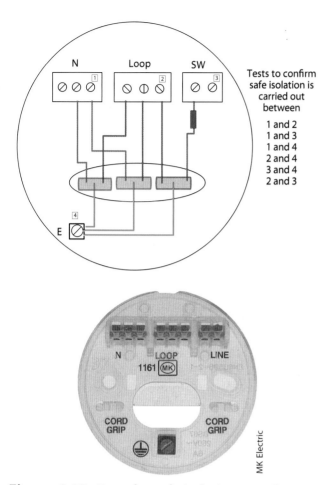

Tests to confirm safe isolation is carried out between

1 and 2
1 and 3
1 and 4
2 and 4
3 and 4
2 and 3

Figure 1.15 *Tests for safe isolation at ceiling rose*

Finally, we must confirm that the approved voltage indicator is still working. To do this we will use the proving unit again and all the LED indicators should light when the approved voltage indicator is tested.

This process will have confirmed that the circuit we are going to work on is isolated from the supply. However, it is always advisable when removing accessories to carry out work to further check at each accessory that isolation is achieved. It may be that a luminaire or accessory which appears to be on the circuit is actually supplied from elsewhere.

For example, it is not uncommon to find a downstairs socket wired from an upstairs circuit and once the downstairs circuit is isolated it would be logical to assume that all the ground floor sockets are isolated. Similarly, where two circuits supply the same area, such as the ground floor of a dwelling, then it is not always obvious which sockets are supplied on which circuit.

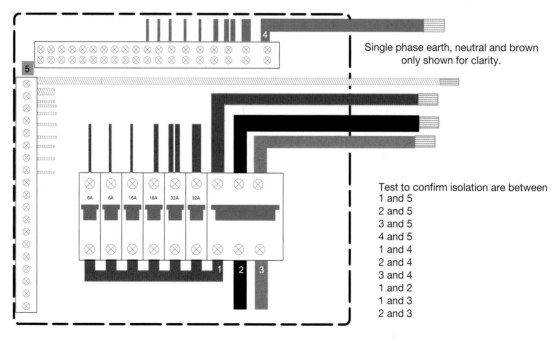

Single phase earth, neutral and brown only shown for clarity.

Test to confirm isolation are between
1 and 5
2 and 5
3 and 5
4 and 5
1 and 4
2 and 4
3 and 4
1 and 2
1 and 3
2 and 3

Figure 1.16 *Three phase points of test for isolation*

Summary procedure for circuit isolation:

1 Seek permission to isolate
2 Identify the circuit to be isolated
3 Isolate by switching off circuit breaker or isolator
4 Fit locking device and warning label
5 Secure area around accessory to be re-moved (barriers)
6 Remove accessory
7 Select an approved voltage indicator (AVI)
8 Confirm AVI complies with GS 38
9 Confirm the operation of the AVI using a proving unit
10 Test between all live conductors
11 Is circuit dead? If not go back to 2
12 Test between all live conductors and earth
13 Is circuit dead? If not go back to 2
14 Confirm AVI is functioning using proving unit.

Isolation of a complete installation follows a similar procedure but in this case we are isolating the supply to the whole installation or all the circuits supplied from a particular distribution board.

Figure 1.17 *Make sure the duty manager is aware of all of the requirements*

When isolating a number of circuits it is important to discuss with the duty holder the areas that will be isolated and determine any special requirements related to any of the circuits which will be isolated.

If we are to confirm safe isolation of a three phase distribution board then there will be a number of additional tests to be carried out as identified in Figure 1.16 on page 11.

Task

It is important that electricians are aware of the kinds of defect which may occur in test equipment. State the three examples of such defects as given in GS 38:

a _____

_____

b _____

_____

c _____

_____

Summary procedure for distribution board isolation:

1 Seek permission to isolate
2 Identify the distribution board to be isolated
3 Isolate by switching off main isolator
4 Fit locking device and warning label
5 Remove distribution board cover to access live terminals
6 Select an approved voltage indicator (AVI)
7 Confirm AVI complies with GS 38
8 Confirm the operation of the AVI using a proving unit
9 Test between all live conductors
10 Is circuit dead? If not go back to 2
11 Test between all live conductors and earth
12 Is circuit dead? If not go back to 2
13 Confirm AVI is functioning using proving unit.

Note

A safe isolation flow chart is included at thetdend of this chapter for your reference. You can copy this and put it with your test equipment as an aide memoire.

Remember

All items of test equipment, including those items issued on a personal basis, should receive a regular inspection and, where necessary, a test by a competent person.

When we consider proving the operation of the approved voltage indicator (AVI) we can use a known live supply or a proving unit. When isolating a distribution board it is possible to use the incoming supply to the isolator to prove the AVI is functioning before and after we test for isolation. This is not possible when isolating a single circuit and so a proving unit is essential for circuit isolation.

Congratulations, you have now completed Chapter 1 so correctly complete the following self assessment questions before you carry on to the next chapter.

SELF ASSESSMENT

Circle the correct answers.

1   Consider the following:

   i.   An isolator must cut off the supply from every source of electrical energy

   ii. A light dimmer switch can be used as a means of isolation

   Which of the following statements is correct in respect of the two statements made above?

   a.   only statement (i) is correct

   b.   only statement (ii) is correct

   c.   both statements are correct

   d.   neither statement is correct

2   'Electrical test equipment for use by electricians' is the title of:

   a.   BS 67

   b.   GS 83

   c.   GS 38

   d.   BS 88

3   The accepted minimum shock current level that could produce ventricular fibrillation of the heart is approximately:

   a.   0.5mA

   b.   0.05A

   c.   0.5A

   d.   5A

4   A three phase and neutral circuit is to be isolated from the supply. The number of tests that will need to be undertaken to confirm safe isolation is:

   a.   3

   b.   6

   c.   7

   d.   10

5   Two electricians are working on different electrical circuits, fed from the same distribution board. To ensure the safety of all parties the isolation of the distribution board's main isolator will require:

   a.   One padlock, a multi-lock hasp and a warning notice

   b.   One padlock, a single lock hasp and two warning notices

   c.   Two padlocks, multi-lock hasp and two warning notices

   d.   Two padlocks, a single lock hasp and a warning notice

Safe isolation flow chart

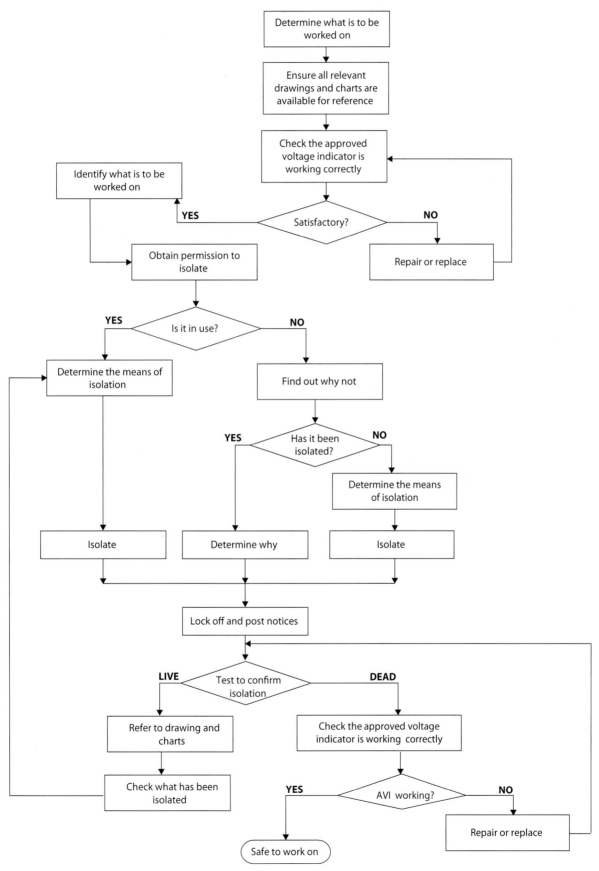

2 Regulatory requirements

RECAP

Before you start work on this chapter, complete the exercise below to ensure that you remember what you learned earlier.

- If the voltage on the equipment to be tested exceeds _____ ac or _____, then test probes and leads should meet the requirements of the _____ and _____ Executive Guidance Note _____.

- To comply with GS 38, test probes should have not _____ than _____ of _____ metal tip and where practicable _____ together with _____ guards to protect against inadvertent hand _____ with _____ parts whilst testing.

- Safe isolation involves _____ the supply is _____ off and it is _____ so that it is not _____ re-energized.

- Failure to _____ isolate when working on electrical _____ and equipment means a _____ risk of electric _____ to ourselves and _____.

- Electric shock also carries the danger of electrical _____ which occur at the _____ and _____ points of the _____ and _____ the body along the _____ taken by the current.

- Before carrying out safe isolation, the effect this action will have on _____, the building and the _____ and _____ must be considered.

To confirm that a three phase circuit is safely isolated, tests are carried out between:

- All _____ conductors and _____
- All _____ conductors and _____
- _____ and _____
- All _____ conductors

and there should be no _____ present at _____ of these connections.

LEARNING OBJECTIVES

On completion of this chapter you should be able to:

- Identify and interpret appropriate sources of relevant information for the termination and connection of conductors, cables and flexible cords in electrical wiring systems and equipment including:

  - Statutory documents and codes of practice

  - British Standards

  - IEE wiring regulations

  - Manufacturers' instructions

  - Installation specifications.

- Specify organizational procedures for reporting variations to the installation specification.

This chapter considers the appropriate sources of information and reporting variations to specifications. You will need to have access to BS 7671 and other guidance material when working through this chapter. It would also be helpful to be able to refer to IET Guidance Note 1 Selection and Erection.

Part 1 Sources of regulatory information

Let's begin by recapping on the regulatory information relevant to electrical installations. This is in two main formats: statutory and non-statutory publications. Statutory requirements place responsibilities on us which are enforceable by law. Non-statutory publications are basically guidance and whilst they do not have a legal status they may be quoted in a court of law in the event of a prosecution under statutory legislation.

The main statutory publications which affect us are:

● The Health and Safety at Work (etc) Act (HSWA)
● The Electricity at Work Regulations (EWR).

The HSWA applies to all work activities and under this statute there are numerous statutory regulations relating to all work activities. It is under this statute that the EWR is placed and this has a direct and specific relevance to our electrical installation work.

The requirements of the statutory documents are quite simple regarding the termination and connection of cables and conductors. In essence, an electrical system must be constructed so that it is safe for use and will not give rise to danger.

Note

Unless exceptional conditions exist there should be no reason for carrying out live working in a consumers' installation.

There is no reason that any process can be valued above the cost of life and so live working can very rarely be justified in any consumers' electrical installation. Carrying out the live tests is one exception and considerable care is required during this process to minimise the risk to the inspector and other people in the vicinity.

In the Electricity at Work Regulations, Regulation 12 refers to electrical connections and states: '*Where necessary to prevent danger, every joint and connection in a system shall be mechanically and electrically suitable for use.*'

More specific information is provided in the non-statutory documents and manufacturers' guidance.

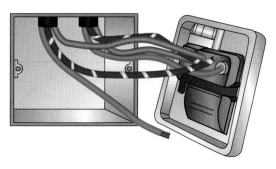

Figure 2.1 *Poor terminations*

BS 7671, Requirements for Electrical Installations, the IET Wiring Regulations published by the Institution of Engineering and Technology, is the principle guidance for the electrical installation industry. Whilst this is not a statutory document it is accepted as standard practice for electrical installation work and may be cited in a court of law.

BS 7671 contains information on the requirements relating to the electrical installation, and the termination of cables and conductors are included in this.

There is a relationship between the requirements of the Electricity at Work Regulations and BS 7671 which is shown in Table 2.1.

We can see from Table 2.1 that many of the requirements of EWR are addressed in BS 7671 but there is no reference for working on or near live conductors. This is because:

● EWR permits working on live conductors providing it is reasonable in all circumstances for the work to be carried out and that appropriate precautions are taken to prevent injury.
● Whilst live testing for confirming compliance and fault-finding may be justifiable, there is no justification for any other work to be carried out live.

Table 2.1 *Comparison of BS 7671 and EWR*

Electricity at Work Regulations	BS 7671
Systems, work activities and protective equipment	Parts 1 and 3
Strength and capability of electrical equipment	Parts 1 and 5
Adverse or hazardous environments	Parts 1 and 7
Insulation placing and protection of conductors	Parts 1 and 5
Earthing or other suitable precautions	Parts 1, 4 and 5
Integrity of reference conductors	Parts 4 and 5
Connections	Parts 1, 5 and 7
Means of protection from excess current	Parts 1 and 4
Means for cutting off the supply and for isolation	Parts 1, 4 and 5
Precautions for work on equipment made dead	Parts 1, 4 and 5
Work on or near live conductors	
Working space, access and lighting	Parts 1, 4, 5 and 7
Persons to be competent and prevent danger and injury	Parts 1 and 2

The BS 7671 numbering system

The numbering system used in BS 7671 may appear a little complex at first and so here is a little guidance as to how it all works.

BS 7671 is laid out in Parts, Chapters and Sections and it is the Parts (Parts 1–7) which form the basis for the numbering of each of the regulations. Each individual regulation is referred to by a unique number.

The first three digits identify, in order, the Part, the Chapter number and the Section of that Chapter.

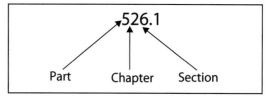

Figure 2.2

So the example in Figure 2.2 is the reference to:

- Part 5 – Selection and Erection of Equipment
- Chapter 52 – Selection and Erection of Wiring Systems
- Section 526 – Electrical Connections.

The final part of the reference defines the particular regulation (1).

This particular regulation is the first in this Section and states:

'526.1 Every connection between conductors or between a conductor and other equipment shall provide durable electrical continuity and adequate mechanical strength and protection.'

In Part 7 Special Locations and Installations the numbering arrangements are different to those used in the other Parts of BS 7671. Part 7 is divided into Sections only so there are no Chapters. Each Section relates to a specific activity and starts with its own section number and then relates the specific regulations, by their number reference, to the main body of the regulations.

Task

Using BS 7671 note the regulation number and the six requirements which must be taken into account when selecting a means of connecting conductors.

Regulation No. _____

Requirements:

1 _____

2 _____

3 _____

4 _____

5 _____

_____

6 _____

Task

BS 7671 identifies exceptions to the requirement that every connection and joint shall be accessible for inspection, testing and maintenance.

1   Identify the Regulation which lists these exceptions.

_____

2   List three of these exceptions.

_____

_____

_____

Example 1: Section 701 Locations Containing a Bath or Shower.

- 701.1 Outlines the scope of Section 701 and the '.1' relates to Part 1 of BS 7671 which deals with the scope of the Standard.
- 701.3 Relates back to Part 3 Assessment of General Characteristics.

- 701.32 Relates back to Chapter Classification of External Influences.
- 701.32.1 Begins the regulation group.

So we can see the numbering in Part 7 is different from that in the main body of BS 7671.

BS 7671 is a harmonized document and the European Electrical Standards Body (CENELEC) numbering system has been adopted to make it easier to accommodate future IEC changes. The Standard includes a unique 120 number system which is used for UK regulations only. One example of this can be found in Section 525, Voltage Drop in Consumers Installations. The first regulation in this section is numbered 525.1 which indicates it is a harmonized requirement. The next three regulations are numbered 525.120 to 525.122 and these regulations are unique to the UK. So wherever we find regulations with the 120 number these are particular to the UK.

Finding tables in BS 7671

Tables in the body of the regulations are numbered using the chapter number and the relative number of tables within that chapter. For example, Table 41.1 is the first table in Chapter 41 and so on.

Figure 2.3 *Make sure you are familiar with the regulations*

> **Remember**
>
> Regulations are numbered and located by Part, Chapter and Section.
> - Tables which have two numbers in their identifier (41.1) are in the Chapter represented by those two numbers.
> - Tables in the Appendices have one number and alphabetical notation, the number indicating the Appendix in which the table can be found.
> - Regulations with a 120 number are unique to the UK.

Tables in the Appendices of BS 7671 are numbered by the appendix number and then the relative alphabetical location of the table within the appendix. For example, Table 3A is the first Table in Appendix A.

In addition to BS 7671 the IET produces other guidance material in the form of Guidance Notes.

These are designed to provide further guidance on the requirements of BS 7671 and include:

- Guidance Note 1 – Selection and Erection
- Guidance Note 2 – Isolation and Switching
- Guidance Note 3 – Inspection and Testing
- Guidance Note 4 – Protection Against Fire
- Guidance Note 5 – Protection Against Shock
- Guidance Note 6 – Protection Against Overcurrent
- Guidance Note 7 – Special Locations
- Guidance Note 8 – Earthing and Bonding.

As you would expect, there is information and guidance regarding the termination of cables and conductors in Guidance Note 1, Selection and Erection.

There are a number of other British Standards and Codes of Practice which may apply to the termination of cables and conductors and these include:

- BS 67: Specification for ceiling roses
- BS 95: Electrical earthing. Clamps for earthing and bonding
- BS 1363: 13A plugs, socket outlets, connection units and adaptors

- BS 5733: Specification for general requirements for electrical accessories
- BS 7909: Code of practice for temporary electrical systems for entertainment and related purposes
- BS 8488: Prefabricated wiring systems intended for permanent connection in fixed installations
- BS EN 50127: Signs and luminous-discharge-tube installations operating from a no-load rated output voltage exceeding 1kV but not exceeding 12kV
- BS EN 50174: Information technology – Cabling installation
- BS EN 50281: Electrical apparatus for use in the presence of combustible dust
- BS EN 60079: Electrical apparatus for explosive gas atmospheres
- BS EN 60309: Plugs, socket outlets and couplers for industrial purposes
- BS EN 60335: Household and similar electrical appliances
- BS EN 60598: Luminaires
- BS EN 60669: Switches for household and similar fixed electrical installations
- BS EN 60947: Low-voltage switchgear and control gear.

This list is not exhaustive and there are many BS and BS EN Standards which have information and requirements for terminations. Some are quite detailed and some are quite general.

There are also CENELEC harmonized documents and the requirements from many of these are incorporated into BS 7671.

Approved Codes of Practice are non-statutory but may be cited in a court of law if such a code of practice is relevant to a matter of prosecution. The most common code of practice used in connection with electrical equipment is the Code of Practice for In-service Inspection and Testing of Electrical Equipment. This Code of Practice does include requirements for connections, in particular those associated with plugs.

Figure 2.4 *A court action could be brought for breach of contract*

Manufacturers' information is generally provided with electrical equipment and accessories. These instructions often refer to the requirements for connection of the equipment. This may include the correct connection for the equipment to function correctly and any special requirements such as torque settings, special encapsulation or sealing procedures and the requirements for enclosures.

Installation specifications may provide specific requirements for the termination of cables and conductors. These requirements may be in excess of those required by the non-statutory documents and manufacturer's instructions. In these instances the specification will take precedence over the other requirements. Should there be a conflict between the manufacturer's instructions and the specification this should be raised with the specifier to resolve the issue.

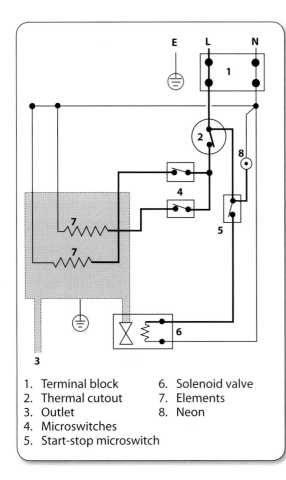

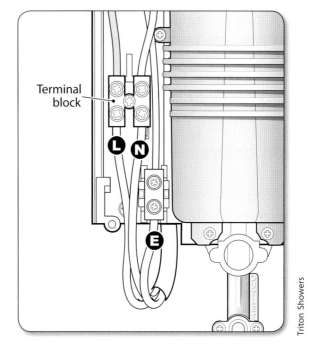

1. Terminal block   6. Solenoid valve
2. Thermal cutout   7. Elements
3. Outlet   8. Neon
4. Microswitches
5. Start-stop microswitch

Triton Showers

Figure 2.5 *Typical manufacturer's instruction for termination*

Task

Using the information in Appendix 1 of BS 7671 state the relevant British Standard for each of the following:

1 A shaver socket outlet installed in a bathroom _____

2 13A switched socket _____

3 A three plate ceiling rose _____

4 The fuse fitted in a 13A plug _____

5 An Edison Screw Lampholder _____

Part 2

Changes to the specification

Changes to the specification for the work being carried out may be generated by the client, the consultant or us, the contractor. Any change to the specification will require documentation to confirm the change has been agreed. This normally involves the issue of a variation order.

Variation orders

Variation orders, sometimes called architect's instructions, are issued when any changes to the original specification are required. On larger contracts a client who wishes to make an alteration would discuss the change with the consultant who would then issue a variation order to the main contractor. The main contractor would then forward the variation order to the relevant subcontractor.

Figure 2.6 *The architect*

Figure 2.7 *The consultant*

Figure 2.8 *The main contractor*

On smaller projects the client will often approach you directly with a verbal request for any changes to the original work. In such instances you should discuss these changes with the client and determine exactly what is required. You should confirm with the client that the information you have recorded is correct and advise them of any possible effect on the works programme as a result.

You should advise whether the work requested is possible and inform the client that they will receive a quotation from your company for the additional work. The information you have recorded should then be forwarded to your office in order for a quotation to be raised for the additional work.

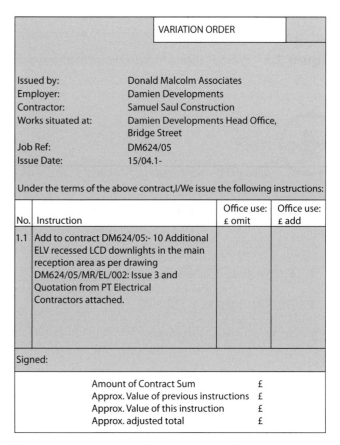

Figure 2.9 *A typical variation order*

The site manager may approach you to carry out some modification not related to the actual installation, for example, a change to the delivery of an item of equipment to site. You would need to record the details and confirm they are correct and advise that the change will require a variation order from the main contractor. Forwarding the information to your company office, together with an estimate of the time and any materials involved, will allow them to prepare a quotation for the work and give advice on the effect to the programme.

Where your company employs subcontractors, you may be involved in asking them to carry out additional or alteration work. A verbal request would normally be the first step and this would be followed by raising a variation order for the subcontractor. You would expect to get a quotation from the subcontractor before authorizing the work.

There may be occasions where you need to request a change to the specification such as a change to an item of equipment. A change may offer a benefit such as a shorter delivery time or reduced cost without loss of quality. A request will need to be made to the client and consultant often through the main contractor. The request for a variation order will need to include a justification for the change identifying the advantages, cost implications and effect on the programme.

If the proposal is accepted a variation order will be issued to confirm the change to the specification.

Most companies have their own format of variation request and variation order but they all contain similar information. The process of dealing with variations to the specification may also vary from company to company. It is common for the main contractor or consultant to impose their forms of documentation and procedures on a contract.

Variation Request			
PT Electrical Contractors Ltd.			
Contract Ref.	DM624/05	Contract address	Damien Developments Head Office, Bridge Street, Walkingham, WK12 9SA
Date:	17/04/1-		
Drawing Ref.	DM624/05/MR/EL/002		
Request	Change of specified ELV recessed LED downlights from order DM624/05/MR/EL/002: 27. Change of manufacturer to Brightlite model LED/WW/10W. (catalogue number 455-731). Specification and standard identical to original Lightwell LEDw10w.		
Costing	Delete item: Lightwell LEDw10w	£***	Replace Item: Brightlite model LED/WW/10W £***
Programme	Improved delivery from 10 weeks to 3 weeks		

Figure 2.10 *Typical variation request form*

Supposing we need to seek a variation to our work where, for example, we need to divert our services around a particular building obstruction such as a supporting beam which was not identified on the drawings.

It is important that the instruction, when it is issued, describes precisely what we are required to do. If it is possible, depending on the urgency of the operation, we should include the cost and programme implications of the request when it is submitted.

When requesting information or instruction it is important to state the latest date by which this needs to be received if we are to remain on programme. This must take account of ordering, delivery and installation constraints.

If a response is not received by that date, we shall need to notify the client of a delay to the programme. This will need to be monitored and the client advised on a regular basis if we are not to incur additional penalties for delay.

We must maintain a register of our requests and this should contain a minimum of:

- Reference number for the request
- Date the request was made
- Latest date for a response
- Subject of the request
- Date response received
- Is the response satisfactory (Yes/No)
- Further action or request made (with the appropriate reference number).

Confirmation of receipt of variation

When any variation is received, we must notify the party who issued the variation order of its receipt and whether it is accepted. There may be occasions where a variation will be questioned due to the need for further information, clarification or for practical reasons.

This confirmation is usually carried out as a matter of course by the company that receives the variation order when it is in the form of an official written instruction. This is because there is normally a need to respond within the stated time period specified in the contract. Many companies have a standard form for this and a typical example is shown in Figure 2.11.

Upon receipt of a variation order on site we must ensure that we record the instruction number, the date it was received and a brief description of the content. We must then investigate what is required and advise the client, within the stipulated time period, of any financial or programme implications.

When we respond to an instruction in this way we must make it clear that this response is only an estimate of the final cost. It will depend on

PT Electrical Contractors Ltd.

To: Contract No.
Client Ref:
Date:

Dear Sirs,

Re: CONFIRMATION OF RECEIPT OF INSTRUCTION
CONTRACT:

We acknowledge receipt from your representative ....
of instructions to vary our works as described below:

We confirm that in accordance with the terms and conditions of the contract between (*Co. name* ....) and (*Client name* ....) the above identified/described instruction(s) which have been received by (*Co. name* ....) will be valued in accordance with the terms and conditions of the contract. The above instruction may result in a variation to our contract price and programme.

Signature of Client's rep.
Issued by:

Figure 2.11 *Confirmation of receipt of instruction*

the nature and extent of the variation as to how accurate our estimate is going to be.

However, there are some occasions when this does not or cannot happen, particularly where instructions are given other than through the official formal channels.

Instructions may be received from the main contractor's representatives on site. Where the variation is of an urgent nature there is usually a procedure in place for the main contractor to issue written instructions on site.

These normally take the form of site-issued, hand-written instructions, but in many cases we will be asked verbally to carry out some work. We must notify the receipt of verbal instructions and confirm exactly what it is we have been asked to do.

Task

The 20 recessed LED luminaires specified as part of a contract are supplied with heat resistant tails but without terminal boxes. The terminal boxes must contain the connections between the luminaire and supply cable and provide clamps to secure the cables. Using manufacturers' information select a suitable terminal box for this task and complete the variation request form in Figure 2.12 for the main contractor for the supply of the boxes selected.

Variation Request			
PT Electrical Contractors Ltd.			
Contract Ref.	DM624/05	Contract address	Damien Developments Head Office, Bridge Street, Walkingham, WK12 9SA
Date:			
Drawing Ref.	DM624/05/MR/EL/002		
Request			
Costing	Delete item: none	£0.0	Replace Item:	£
Programme			

Figure 2.12 *Variation order*

This site-issued instruction will normally be followed by a variation order for the work described in the site instruction. Where this is not forthcoming a formal request needs to be made for a variation order together with a copy of the site instruction.

Confirmation of the receipt of variation instructions is to be sent, irrespective of how the instruction is received or from whom. Many contracts state that if notification of receipt is not refuted, usually within 14 days, then it is deemed that the instruction has been accepted and we are entitled to be paid for the work involved.

Remember

A variation order may not carry any cost or programme implication but without it we will be obliged to install to the original requirements. If the variation order is not issued then we will not be paid for any uninstructed variations.

Try this

Complete the wordsearch and identify the word which is not included in the puzzle.

Missing word: _____

```
C I N S T R U C T I O N S O E
O A X S K A P P E N D I X S N
N R E G U L A T I O N S N O S
F H A R M O N I Z E D O I D G
I S T A T U T E N X P T R N R
R E I X B D E H U S A A O O E
M C Y C G E E A E L D I T W Q
A T L L U P I R L N T C P R U
T I E I I F K A A A A O E C E
I O G E D B T T I R L T E P S
O N A N A S S R T A P F K A T
N H L T N F A N S A H W S R Z
S Z S I C V O A H I Y M A T Z
Y Z H W E C Z C E N E L E C I
F F C M A D D I T I O N A L K
```

Additional	Harmonized	Cenelec	Instructions	Chapter
Regulations	HSWA	Request	Confirmation	Appendix
Contractor	Order	Guidance	Client	Installation
Statute	Variation	Legal	Standards	Part
Response	Section			

Congratulations you have now completed Chapter 2 of this study book. Complete the self assessment questions before continuing to Chapter 3.

SELF ASSESSMENT

Circle the correct answers.

1   The statutory document which identifies a requirement for the quality of terminations is:
   a.   BS 7671
   b.   The Building Regulations
   c.   Electricity at Work Regulations
   d.   Health and Safety at Work (etc) Act

2   The Figure 5 in BS 7671 Regulation 526.1 indicates that the regulation can be found in:
   a.   Part 5
   b.   Section 5
   c.   Chapter 5
   d.   Regulation 5

3   An Approved Code of Practice may be cited in a court of law if it:
   a.   Has been in place for more than 5 years
   b.   Is relevant to a matter of prosecution
   c.   Does not aid the defence
   d.   Has statutory status

4   Changes to the technical specification for an electrical installation may be notified to a subcontractor using a:
   a.   Time sheet
   b.   Works order
   c.   Daywork sheet
   d.   Variation order

5   A request is to be made by the electrical contractor for a change to the specification for an item of equipment. Which of the following need **not** be included in the request?
   a.   The date by which a response is required
   b.   The programme implications
   c.   The name of the wholesaler
   d.   The cost implications

RECAP

Before you start work on this chapter, complete the exercise below to ensure that you remember what you learned earlier.

- Statutory documents are _____ by _____ whilst non-statutory documents may be _____ in a _____ of law.

- The Health and Safety at Work (etc) Act applies to _____ work activities and under this statute there are numerous _____ regulations.

- Regulation numbers in BS 7671 are arranged in order: the _____, the _____ number and the _____.

- IET Guidance Note 1 is a _____ document providing guidance on the _____ and _____ of electrical equipment and _____.

- Manufacturers' _____ often refer to the requirements for the correct _____ of the electrical equipment including any special requirements such as _____ settings.

- _____ orders are issued when any _____ to the original _____ are required.

- Sending the information on any changes to your company office, together with an _____ of the _____ and any _____ involved, will allow them to prepare a _____ for the work and advise on any effect on the _____.

- When requesting information or _____ it is important to state the _____ date by which this needs to be received if we are to _____ on programme.

- When any _____ is received we must _____ receipt and whether it is _____ .

LEARNING OBJECTIVES

On completion of this chapter you should be able to:

- Describe methods and procedures appropriate to the installation environment to ensure the safe and effective termination and connection of conductors, cables and flexible cords in electrical wiring systems and equipment including:

  - Thermosetting insulated cables

  - Single and multicore pvc and thermosetting insulated cables

  - pvc/pvc flat profile cable

  - MICC (with and without pvc sheath)

  - SWA cables (PILC, XLPE and pvc)

  - Armoured/braided flexible cables and cords

  - Data cable

  - Fibre optic cable

  - Fire resistant cable.

This chapter considers the methods and procedures appropriate to the installation environment to ensure the safe and effective termination and connection of conductors, cables and flexible cords in electrical wiring systems.

Whilst working through this chapter you will need to refer to:

- BS 7671

- IET Guidance Note 1 Selection and Erection

- Manufacturers' information

- IET On-Site Guide may also be helpful.

Part 1 The requirements for connections

The termination and connection of cables and conductors form an everyday part of an electrician's work and these connections must meet the requirements of BS 7671 for safety. To meet these requirements we need to understand what they are and select the most appropriate method in each case.

We first need to understand what is required of a termination or connection. We can then consider the environment in which the termination is to be made and which will be present during its normal operation.

BS 7671 sets out the basic requirement for all connections in Chapter 52 and we need to be familiar with them.

When terminating or connecting cables or conductors one basic requirement is that every cable or conductor is supported so that it is not subjected to undue mechanical strain and that there is no *appreciable mechanical strain on the termination of the conductors'*. This includes any strain caused by the supported weight of the conductor or cable itself.

This requirement results in the need for cable clamps or supports in any item of equipment where strain could be placed upon the terminations such as we see in the basic 13A plug or ceiling rose.

As far as the actual connections are concerned there are a number of points to consider.

Section 526 of BS 7671 identifies the particular requirements for the connections and terminations which include the following.

Every connection must be accessible for inspection, testing and maintenance. The only exceptions

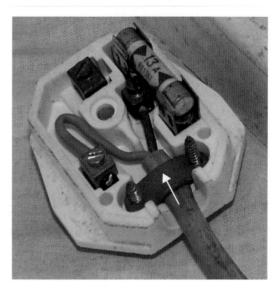

Figure 3.1 *A 13A plug supplying Class II equipment with the cord-grip indicated*

to this requirement are specific methods of connection including:

- A joint made by a suitable compression tool
- A joint designed to be buried in the ground
- A compound filled or encapsulated joint
- A joint made by soldering, brazing or welding
- Maintenance free (MF) accessories complying with BS 5733, marked with the symbol (MF) and installed in accordance with the manufacturer's instructions
- A connection between the cold tail and the heating element of a heating system such as underfloor heating
- Joints in equipment made by a manufacturer which are not intended to be inspected or maintained.

This means that a screw terminal junction box beneath a wooden floor in a dwelling does not meet the requirements and must be accessible for maintenance and testing.

Kindly reproduced by permission from Hager Ltd

Figure 3.2 *Screw terminal joint box must be accessible but maintenance free joint box does not*

The protection of the termination or connection is also to be considered and the termination of live conductors (line and neutral) must provide basic protection. This is achieved by using one, or a combination, of the following:

● A suitable accessory which complies with an appropriate product standard

● An equipment enclosure complying with an appropriate product standard

● An enclosure formed by building material which is non-combustible (tested to BS 76-4).

There is also a requirement to ensure that where a connection is made within an enclosure that the enclosure provides protection against any external influence and offers adequate mechanical protection for the connection.

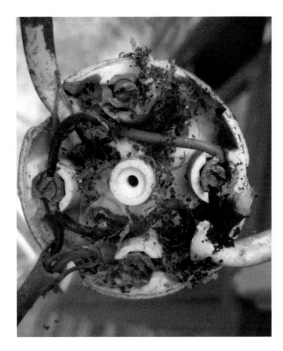

Figure 3.3 *Not a suitable enclosure for the location*

This means that a connector strip and insulation tape is not a suitable connection method for a recessed luminaire. There is no mechanical protection provided and the accumulation of dust and debris on the terminals creates a risk of arcing and fire.

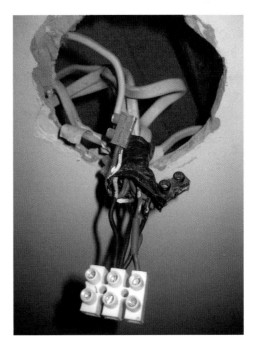

Figure 3.4 *Not a suitable method of terminating live conductors*

To meet the requirements of security of connection, accessibility and mechanical protection, a suitable enclosure will be required. There are many products on the market which provide a suitable enclosure and one such enclosure, suitable for the connection of a luminaire in a dwelling is shown in Figure 3.5. This provides cable clamps to secure the cables, an enclosure to contain and protect the connections and the completed enclosure provides protection against the external influences likely to be encountered.

Figure 3.6 *Failure to suitably locate the enclosure*

Kindly reproduced by permission from Hager Ltd

Figure 3.5 *A typical enclosure for a domestic luminaire connection*

Once the termination is completed the enclosure should be located so that it provides and continues to provide suitable protection for the terminations. The location of the enclosure should also take account of any external influences and a common problem with downlights is the generation of heat.

The location of the enclosure should take account of any external influence, such as the generation of heat, which is a common problem with downlights. Figure 3.6 shows the result of the failure to do so. The material of the enclosure is not flammable which means that the heat has not ignited the material, however the damage to the enclosure is evident and the enclosure has been melted through and the connections inside subjected to considerable heat.

Where cables and conductors enter an enclosure they should be suitably protected against damage. Where cables and conductors enter enclosures there should be suitable protection between the enclosure and the cables and conductors. Where single core cables, or where the sheath of sheathed cables is removed, the conductors should be suitably protected against damage.

For example, where a sheathed cable enters a metallic enclosure there must be some protection against the cable being damaged by contact with the metal. This may be achieved in a number of ways including:

● The use of pvc or rubber grommets
● Bushes and couplers or locknuts
● Stuffing glands
● Grommet strip.

Failure to provide this protection means that the cables are liable to damage and the requirement to protect against external influences must also be considered.

Task

Familiarize yourself with the requirements of BS 7671, Section 526 Electrical Connections before you continue with this chapter.

Steel boxes should be fitted with a grommet so that no sharp edges can damage the cable.

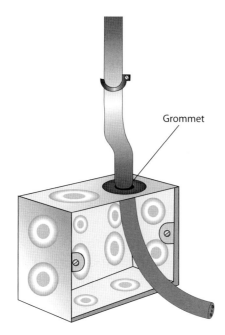

Figure 3.7 *Cable entering a steel box*

A plastic pattress with 'knockout' sections should have the section removed so that there are no sharp edges and the hole is the same size as the cable.

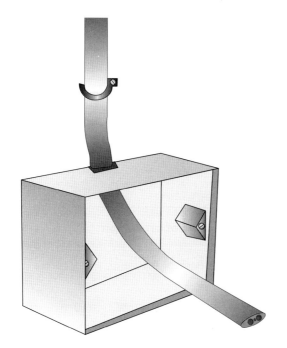

Figure 3.8 *Cable entering a plastic pattress*

This is best carried out using a junior hacksaw to determine the width of the hole and a pair of pliers to remove the plastic between the two cuts.

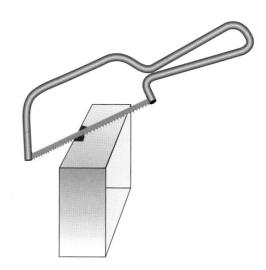

Figure 3.9 *Cut the cable entry on a plastic pattress*

Where enclosures or barriers are used to provide basic protection there are a number of important issues and some of these relate to the ability of the enclosure to provide protection from external influences.

Live parts (which include terminations and connections) must be inside an enclosure or behind barriers which provide protection to at least IPXXB or IP$_2$X. The enclosure or barrier must be securely fixed and in addition the top surface of an enclosure which is readily accessible must provide a level of protection at least equal to IPXXD or IP$_4$X.

The use of the IP Code to identify the level of protection afforded by barriers and enclosures is considered in detail in the Planning and Selection for Electrical Systems study book. Table 3.1 shows the basic IP Code.

Table 3.1 *Basic IP Code*

IP Code				
1st Digit	**Level of protection**	**2nd Digit**	**Level of protection**	
0	Not protected	0	Not protected	
1	Protected against solid foreign objects of 50mm diameter and greater	1	Protected against vertically falling water drops	
2	Protected against solid foreign objects of 12.5mm diameter and greater	2	Protected against vertically falling water drops when enclosure is tilted up to 15°	
3	Protected against solid foreign objects of 2.5mm diameter and greater	3	Protected against water sprayed at an angle up to 60° on either side of the vertical	
4	Protected against solid foreign objects of 1.0mm diameter and greater	4	Protected against water splashed against the component from any direction	
5	Protected from the amount of dust that would interfere with normal operation	5	Protected against water projected in jets from any direction	
6	Dust tight	6	Protected against water projected in powerful jets from any direction and heavy seas	
No code		7	Protected against temporary immersion in water	
No code		8	Protected against continuous immersion in water, or as specified by the user	
Additional letters				
Additional letter		**Level of protection**		
A		Back of the hand		
B		Finger (12mm)		
C		Tool (2.5mm)		
D		Wire (1.0mm)		

Remember

The additional letters, such as IPXXB, are used to identify levels of protection against access to live parts by persons.

The cable entry into the switch fuse in Figure 3.10 does not meet any of the requirements for the cable entry, protection against damage or environmental conditions.

It should only be possible to access the enclosure or remove a barrier:

- By the use of a key or tool
- Only after disconnection of the supply to the live parts within

- Where an intermediate barrier to at least IPXXB or IP2X is installed which can only be removed by the use of a key or tool.

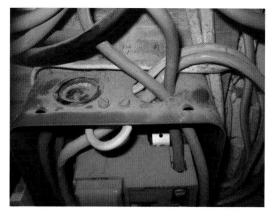

Figure 3.10 *Cable entry does not meet the requirements for enclosures or terminations*

Task

Familiarize yourself with the requirements of BS 7671, 416.2 Barriers and Enclosures before you continue with this chapter.

Try this

1 State the most suitable method of protection for a cable entering an enclosure for the following situations:

a A pvc sheathed flat cable entering a metal switch box from below.

_____

b A number of single core cables contained in a steel conduit entering a metal distribution board.

_____

c A set of meter tails entering an insulated consumer unit from below.

_____

d A steel wire armoured (SWA) cable terminating at a metallic motor starter.

_____

2 State the level of IP protection which needs to be provided for:

a A cable entering a surface socket outlet box from above.

_____

b The front cover of a consumer unit.

_____

c The internal barrier contained within a metal clad distribution board.

_____ .

Part 2 The connections

Having considered the cable entry and enclosures we will now consider the actual requirements for the connections. The first basic requirement is that every connection between conductors, and between conductors and equipment, is mechanically and electrically sound. So they must provide good electrical continuity and mechanical strength.

Any termination of an insulated conductor, other than a protective conductor $\geq 4mm^2$, should be made in an accessory or luminaire complying with the appropriate British Standard.

In order to ensure that the termination meets these requirements there are a number of considerations to be made. These include making sure the terminations are compatible with:

- The cross-sectional area (csa) of the conductor
- The shape and number of wires which make up the conductor
- The number of conductors to be connected together
- The temperature the terminals will reach in normal service
- Suitable locking arrangements where vibration or thermal cycling is likely.

The cross sectional area of the conductor(s) to be terminated determines the physical size of the terminal to be used. Where a conductor is to be terminated the terminal must be the correct size for the termination to be both electrically and mechanically sound.

The shape and number of strands which make up the conductor will affect the termination to be made. All the strands of stranded conductors must be contained within the termination. The csa and hence the current carrying capacity of the conductor has been carefully selected, any strands not contained within the termination reduces the current carrying capacity and may result in the termination becoming too hot with a risk of damage to the insulation and subsequent fire. Solid shaped conductors are to be terminated in a suitably shaped terminal; this is generally where cables are crimped to a lug or terminal pin for connection into switchgear and equipment.

Figure 3.11 *Shaped conductor lug*

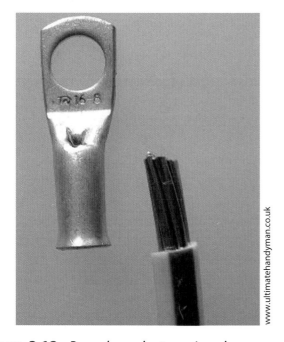

Figure 3.12 *Round conductor crimp lug*

The capacity of the termination must be suitable for the number of conductors to be connected. In many accessories this is achieved by having multi-way terminals such as the loop-in terminal in a ceiling rose. Where more than one conductor enters a terminal it is essential that the terminal can accommodate all the conductors and provide a sound termination both electrically and mechanically. Cable glands, clamps and compression-type joints should retain all the strands of the conductor securely.

The operating temperature of conductors with thermoplastic insulation is 70°C and the majority of electrical accessories are designed to operate safely with the conductors at this temperature. Thermosetting cables may operate at 90°C with specialist cables operating at much higher temperatures. It is important to confirm that the terminals and terminations to be used are suitable for the temperatures they will reach in normal service.

Vibration and thermal cycling may affect the quality of the termination and these need to be considered. Where there is vibration this can reduce the mechanical security of the termination over time, particularly where the conductors are in free air. Thermal cycling is where the nature of the connected load causes current to flow for a period of time, then stop flowing, then flow again on a repeated cycle. This results in heating and cooling of the conductor and the terminals which softens and hardens the material. This happens typically where motor loads are subject to stop–start activity, such as lifts and hoists. By providing a suitable anchor these effects can be considerably reduced.

We also need to consider the corrosion of the termination and connection of the cables and conductors.

www.ultimatehandyman.co.uk

Basically, corrosion is the destruction of metal by a chemical reaction. The most common form of corrosion which we see on a regular basis is the formation of rust on steel which has oxidized due to exposure to the weather. Many of the cable and conductor materials are not ferrous and yet copper plate on roofs changes to green as a result of oxidization. One of the main concerns for our connections and terminals is corrosion due to an electrochemical reaction between the materials and the surrounding environments.

There are three conditions which must all exist at the same time in order for this 'electrolytic' corrosion to take place:

- The presence of an anode and a cathode. This typically occurs when two dissimilar metals are in contact with, or in close proximity to, one another. For example, aluminium and brass
- A metallic connection between the anode and cathode
- An electrolyte such as water.

Dissimilar metals can result in corrosion due to electrolytic action. For example, as brass and aluminium react, a copper termination stud may be used where aluminium conductors are to be connected in brass terminals.

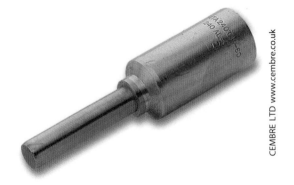

CEMBRE LTD www.cembre.co.uk

Figure 3.13 *Aluminium to copper pin-type termination*

These connections are made in controlled conditions to prevent corrosion between the aluminium and copper. The aluminium tube is normally filled with grease to prevent oxidization of the aluminium. This prevents corrosion within the terminal and ensures a good electrical connection between the aluminium tube and conductor. The conversion from aluminium to copper prevents the electrolytic action between the aluminium and brass. Brass glands are generally used to terminate steel wire armoured (SWA) cables and a similar problem can occur where these are terminated into an aluminium gland plate.

Figure 3.14 *Typical bathroom installation*

The presence of moisture will accelerate the electrolytic process and particular care must be taken where such conditions exist.

The environmental conditions will also need to be considered when carrying out the termination of cables and conductors. It stands to reason that the termination of conductors and cables should be carried out in a suitable environment. Damp and wet conditions can compromise the termination and may result in premature failure. The insulation used in mineral insulated cables is hygroscopic which means it absorbs moisture, so the terminations must be carried out in dry conditions. The armouring of SWA and aluminium

Task

Using manufacturers' information from catalogues or the Internet identify five different types of connection which may be used to terminate aluminium cored cables to a bolted terminal.

1 _____

_____

2 _____

_____

3 _____

_____

4 _____

_____

5 _____

_____

armoured cables may also be affected if they are terminated in damp conditions as this may lead to corrosion of the armour.

The temperature at which cables and conductors are terminated is also a factor to be considered. pvc and thermoplastic cables should not be worked at temperatures below 0°C as they become brittle and may be split and cracked during the termination process. High temperature may also cause problems when terminating these cables as they become more elastic and as a result shrinkage may occur when the temperature drops. This can result in too much bare conductor being exposed, resulting in a risk of fault or electric shock. MIMS cables may require high temperature sealing compound in conditions where high ambient temperatures are likely.

Whilst the appearance of the installation cannot really be classed as an external influence it may play an important part in the choice of the wiring system and the terminations should be

Figure 3.15 *An environment with variable temperatures*

considered as part of this process. The chosen system and method may also affect the type of termination required.

For example, MICC cables are often used in installations of particular significance because of their reliability, longevity and compact size. In a cathedral, for example, these may have to be run on the surface and it is common for bare (unsheathed) cables to be used. These will require surface termination with suitable glands. However, in a concealed MICC wiring system the terminations will not be accessible and so glands would not be appropriate. Special accessory boxes will be required which hold the termination pot securely, and an earth tail pot would be used to provide reliable earth continuity. If a gland is used in these locations the use of the earth tail pot means the gland is simply anchoring the cable and not an electrical connection. Methods of termination are considered in more detail in Chapter 4 of this study book.

Where terminations are to be carried out in locations where dust and foreign bodies may be present, special care needs to be taken. The termination and connection of cables and conductors must be free from foreign bodies to ensure a sound mechanical and electrical connection. Certain cable types, such as fibre optic cables, must be terminated in a clean and dust free environment. The enclosures used to house the terminations and connections must have an IP rating appropriate for the environment and our terminations and connections must not compromise these ratings.

Figure 3.16 *Dusty environment*

Corrosive and polluting substances are sometimes present in the location where the cables and terminations will operate. MIMS cable is available with a stainless steel sheath for installation in areas where corrosion is a high risk. The termination of the cable must also use a stainless steel gland to maintain the protection. In respect of our terminations the enclosure needs to be suitable to provide protection against the substances present. It therefore follows that the connections and terminations should also be appropriate. This may involve the use of additional materials such as compounds and silicone seals in order to protect the electrical terminations.

Failure to use these additional measures can have serious consequences. Figure 3.15 shows the termination to an earth electrode which has been installed in a purpose-built electrode pit but no additional protection has been used for the termination. As a result of the conditions which exist within the electrode pit, the termination and the connection have begun to corrode and action is now required to repair the damage. The use of a corrosion inhibiting grease or compound, together with a plastic shroud once the termination had been made, would have prevented the corrosion and still allowed inspection to take place.

Figure 3.17 *Corroded earth electrode termination*

Mechanical strain was briefly mentioned earlier in this chapter and the requirement for connections and terminations to offer suitable protection is essential. The most common stresses are tensile stress and vibration. To prevent the tensile stress the cable should be suitably anchored to prevent damage to the cable and to ensure that undue stress and strain are not placed on the conductors and terminals.

This is generally achieved by one of the following:

- Cable clamps
- Cable glands
- Containment systems.

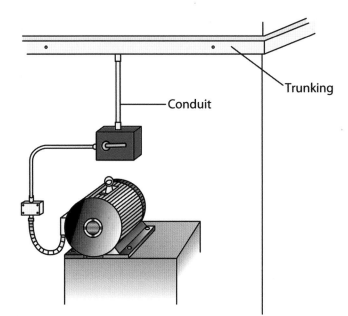

Figure 3.18 *Typical conduitand trunking application*

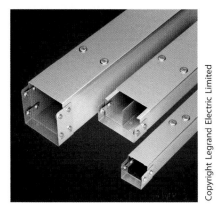

Figure 3.19 *Steel trunking*

Cable clamps and glands secure the cable mechanically and prevent movement. The use of the containment system prevents contact with

the cables and conductors and so stress cannot be placed on them once they are correctly installed.

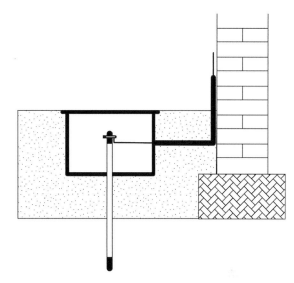

Figure 3.20 *Correctly installed earth electrode*

To prevent vibration affecting the terminations it is common to use either:

- A flexible connection system and cable, or
- Anti-vibration loops in the cable.

The use of flexible cables and flexible conduit are the most common methods for preventing vibration.

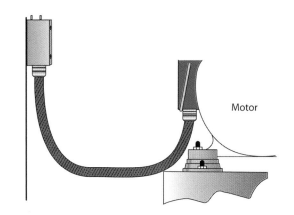

Figure 3.21 *Flexible conduit connection*

The flexible conduit is used where there is a need for additional mechanical protection and single core flexible cables are generally installed within it. Insulated and sheathed flexible cables may be used where the risk of mechanical damage is low, for example, connection to the refrigerator compressor in a dwelling.

Anti-vibration loops are used where the wiring system is not flexible enough to prevent the transfer of vibration and so a loop is included before the termination point. This allows the vibration to be absorbed without undue damage to conductors or terminations.

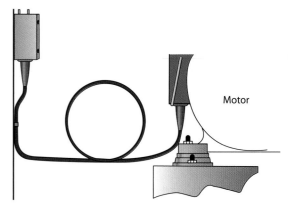

Figure 3.22 *Anti-vibration loop in MIMS cable*

  Try this

State all the environmental hazards likely to be present when cables are to be terminated in the following locations:

1   A motor isolator in a cement processing works.

_____

2   A luminaire in a carwash.

_____

3   A street sign at the seafront.

_____

Congratulations, you have now completed Chapter 3 of this study book. Complete the self assessment questions and the progress check before you continue to Chapter 4.

SELF ASSESSMENT

Circle the correct answers.

1 Every screwed termination must be:
   a. Accessible for maintenance
   b. Contained in a metallic enclosure
   c. Exposed to the air for cooling
   d. Enclosed in epoxy resin

2 The minimum level of protection required for the top surface of an accessible distribution board is:
   a. IP2X
   b. IPXXB
   c. IPXXD
   d. IP4X

3 Where an insulated and sheathed cable enters a metallic enclosure:
   a. There must be protection against mechanical damage
   b. A suitable warning label must be adjacent to the entry
   c. The cores must be sleeved with additional insulation
   d. The sheath must not enter the enclosure

4 Which of the following is **not** a consideration when selecting a termination for a conductor?
   a. The csa of the conductor
   b. The shape of the conductor
   c. The position of the termination
   d. The temperature in normal service

5 An aluminium conductor is to be terminated in a brass terminal using an aluminium to copper crimp lug. This is necessary to:
   a. Prevent corrosion
   b. Allow easy termination
   c. Provide mechanical strength
   d. Identify the conductor rating

Progress check

Tick the correct answer

1. **When carrying out work on an existing circuit the first action to be taken is:**

   ☐ a. safe isolation

   ☐ b. lock off the supply

   ☐ c. obtain permission to isolate

   ☐ d. isolate the supply for the installation

2. **One result of carrying out safe isolation of a circuit is:**

   ☐ a. the circuit will function normally

   ☐ b. it will take longer to complete the work

   ☐ c. part of the installation will not function normally

   ☐ d. there will be no effect on the use of the installation

3. **Failure to carry out safe isolation when working on a circuit will result in:**

   ☐ a. inconvenience to the user

   ☐ b. a shock risk to you and others

   ☐ c. the work taking longer to complete

   ☐ d. the use of the installation will be restricted

4. **When confirming safe isolation the approved voltage indicator will need to be proved:**

   ☐ a. only after use

   ☐ b. only before use

   ☐ c. before and after use

   ☐ d. at each stage of the testing

5. **To confirm safe isolation of a three phase circuit, tests must be carried out between:**

   ☐ a. all line conductors

   ☐ b. all live conductors

   ☐ c. all line conductors and all line conductors and neutral

   ☐ d. all live conductors and all live conductors and earth

6. **Where a number of electricians are working in the same location, the arrangements for ensuring safe isolation will require each electrician to have:**

   ☐ a. a key for a common padlock

   ☐ b. an individual padlock with a common key

   ☐ c. a key for a common padlock and personal warning notices

   ☐ d. their own unique padlock and key separate from all others

7. **One of the considerations given in the Electricity at Work Regulations where work on or near live conductors is to be carried out, is whether:**

   ☐ a. the client is aware the work is to be undertaken

   ☐ b. the DNO has been notified of the work

   ☐ c. the work is confirmed as necessary

   ☐ d. the supply can be isolated

8. **The numeral 2 in BS 7671 Regulation 526.1 indicates that the regulation can be found in:**

☐ a. Part 2

☐ b. Section 2

☐ c. Chapter 2

☐ d. Regulation 2

9. **Approved Codes of Practice are:**

☐ a. non-statutory and may not be cited in a court of law

☐ b. non-statutory but may be cited in a court of law

☐ c. only relevant when stated in a contract

☐ d. statutory and must be complied with

10. **A subcontractor would be notified of changes to the technical specification using a:**

☐ a. variation order

☐ b. daywork sheet

☐ c. works order

☐ d. time sheet

11. **A request to change a specified item of equipment need <u>not</u> include the:**

☐ a. cost implications

☐ b. name of the wholesaler

☐ c. programme implications

☐ d. date by which a response is required

12. **Notification of receipt must be provided for all:**

☐ a. time sheets

☐ b. works orders

☐ c. daywork sheets

☐ d. variation orders

13. **All terminations in conductors and between conductors and equipment must be:**

i) **electrically sound**
ii) **mechanically robust**

☐ a. only statement i) is correct

☐ b. only statement ii) is correct

☐ c. both statement i) and statement ii) are correct

☐ d. neither statement i) nor statement ii) is correct

14. **Which of the following terminations does not have to be accessible?**

☐ a. a bolted terminal in a main control panel

☐ b. a screwed terminal at a luminaire

☐ c. a bonding clamp to a water pipe

☐ d. a crimped joint in a cable

15. **The main tails to a consumer unit enter through a hole in the bottom of the enclosure. The minimum level of ingress protection which must be provided against solid bodies is:**

☐ a. IP1X

☐ b. IP2X

☐ c. IP3X

☐ d. IP4X

16. **The conductors of a flex for a suspended pendant lampholder must be connected so that they are:**

☐ a. solely reliant on the terminals for the support of the lampholder

☐ b. twisted together at the termination point

☐ c. held by a cord grip to prevent strain on the terminals

☐ d. not accessible for testing or maintenance

17. **The connection of dissimilar metals at cable terminations requires care to ensure that corrosion does not occur due to:**

    ☐ a. electrolytic action

    ☐ b. vibration

    ☐ c. cross polarity

    ☐ d. osmosis

18. **Which of the following is <u>not</u> a consideration where a cable enters an enclosure?**

    ☐ a. mechanical damage

    ☐ b. movement of the conductors

    ☐ c. ingress of solids and liquids

    ☐ d. identification of conductors

19. **A crimp lug for a cable termination is to be selected. The factors to be considered are the conductor size, the conductor material and the conductor:**

    ☐ a. length

    ☐ b. location

    ☐ c. shape

    ☐ d. function

20. **An installation has been constructed using metallic conduit and single core cables and the final connection is to be made between a motor starter and the motor terminals. One of the factors which needs to be considered is:**

    ☐ a. dust

    ☐ b. vibration

    ☐ c. water

    ☐ d. damp

Termination of cables and conductors

4

RECAP

Before you start work on this chapter, complete the exercise below to ensure that you remember what you learned earlier.

- When terminating conductors one basic requirement is that there is no _____ mechanical _____ on the _____ of the conductors.

- Accessories complying with BS 5733, marked with the symbol _____ and installed in accordance with the _____ instructions do not need to be _____ for maintenance.

- Once the termination of a conductor is completed the enclosure should be such that it __ and continues to _____ suitable _____ for the terminations.

- Where cables and conductors enter enclosures there should be suitable protection between the _____ and the _____ and _____.

- The IP Code is used to identify the level of protection afforded by _____ and enclosures against the ingress of _____ bodies and _____.

- Any termination of an _____ live conductor should be made in an _____ or _____ complying with the _____ British Standard.

- Terminations must be compatible with the cross-sectional area of the _____, the _____ and number of _____ which make up the conductor and the _____ of conductors to be _____ together.

- _____ metals can result in _____ due to _____ action, for example, brass and _____ react with one another.

- To prevent _____ on terminations, cables should be suitably anchored. This is generally achieved by cable _____, cable _____ or _____ systems.

- To prevent vibration affecting the terminations it is common to use either a _____ connection _____ and cable or _____ loops in the _____.

LEARNING OBJECTIVES

On completion of this chapter you should be able to:

- Explain the termination techniques for:

  - Single and multicore pvc and thermosetting insulated cables

  - pvc/pvc flat profile cable

  - MICC (with and without pvc sheath)

  - SWA cables (PILC, XLPE and pvc)

  - Armoured/braided flexible cables and cords

  - Data cable

  - Fibre optic cable

  - Fire resistant cable.

- Explain the advantages, limitations and applications of the following connection methods:

  - Screw

  - Crimped

  - Soldered

  - Non-screw compression.

- Describe the procedures for proving that terminations and connections are electrically and mechanically sound and explain the consequences if they are not in terms of:

  - High resistance joints

  - Corrosion and erosion.

This chapter considers the methods and procedures to ensure the safe and effective termination and connection of conductors, cables and flexible cords in electrical wiring systems.

Whilst working through this chapter you will need to refer to:

● BS 7671

● IET Guidance Note 1 Selection and Erection

● Manufacturers' information

● IET On-Site Guide may also be helpful.

In the previous chapter we considered the environmental requirements for the termination of conductors and cables. We now need to consider the termination of the various types of cables which are used in electrical installations.

Part 1 Termination techniques for single and multicore thermoplastic and thermosetting insulated cables

The termination techniques are the same for both thermosetting and thermoplastic insulation. We will begin with one of the most common cables, the flat profile multicore cable.

Multicore sheathed wiring

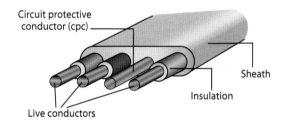

Figure 4.2 *Flat profile three core and cpc insulated and sheathed cable*

Stripping the cable

There are two options for stripping this type of cable and the preferred method involves using a stripping knife. This is not a thin blade trimming knife often referred to as a 'Stanley Knife'. The blade of this type of knife is too

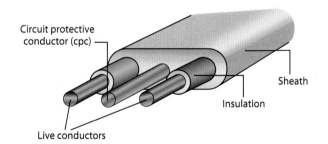

Figure 4.1 *Flat profile twin and cpc insulated and sheathed cable*

thin and too sharp which can easily damage the insulation and increases the risk of injury to you.

There is a bare circuit protective conductor (cpc) in this type of cable and this can be used as a guide for the knife when removing the cable sheath. Allow the knife to lay against the cpc and push slowly with a slight sawing movement. When you have reached the required point peel the sheath back and cut it off with side cutters.

There is nothing wrong with this procedure but it does require practice and a degree of skill. Too much tension on the conductor can stretch the copper and weaken the conductor, and too much pressure will cut through the conductor. Providing care is taken not to damage the cpc this is an acceptable and quick method of removing the sheath.

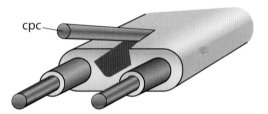

Figure 4.4 *The circuit protective conductor being used to strip the sheath from the cable*

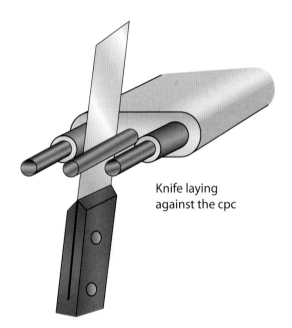

Figure 4.3 *Care must be taken NOT to cut into your hand*

Knife laying against the cpc

Remember

When stripping multicore cables DO NOT:

● Damage the insulation

● Damage the conductor

● Cut the sheath back too far

● Cut the insulation too deep so that you damage the conductor.

Many experienced electricians will not use a knife to strip the sheath off; they will use a pair of side cutters. For this method the cable is first nicked in the end with the side cutters. The cpc is gently gripped and then pulled up through the slot with the side cutters. By continually pulling the cpc back it will cut through the sheath and release the other cores. The sheath is then trimmed off using the side cutters.

The insulation is best removed using cable strippers and there are two main types available. One type has V shaped slots in the blades and these must first be adjusted so that the blade only cuts into the cable insulation and not the conductor. The cutting blades are then placed over the core and squeezed so that the blades cut into the insulation but stop before they reach the conductor. The insulation is now pulled off the conductor.

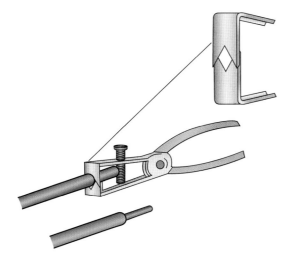

Figure 4.5 *Using cable strippers*

The other popular type is the auto-adjusting type which has jaws made up of a number of spring blades shaped to provide a cutting edge. When these jaws are squeezed onto the cable the spring steel ensures that they automatically adjust to the insulation thickness and do not penetrate far enough to damage the cable. The strippers are then pulled away removing the insulation.

Image courtesy of KNIPEX Germany, see knipex.de

Figure 4.6 *Auto-adjusting strippers*

Of course the insulation can be removed using a knife but care must be taken to avoid cutting your hand, damaging the conductor or damaging the other cores.

When a sheathed cable is being terminated the outer sheath must enter the enclosure and no exposed cores must be left showing. As the cpc is not normally insulated, green and yellow sleeving must be slid over it so that it is insulated throughout its length within equipment or an accessory.

Single core thermosetting and thermoplastic cables are stripped for terminations using the same methods as used for the cores of the flat profile cable cores.

Insulated and sheathed flexible cable

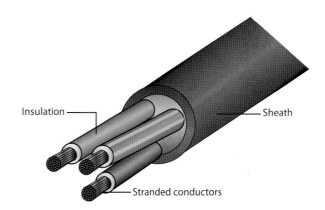

Insulation ——   —— Sheath

—— Stranded conductors

Figure 4.7 *Flexible cable construction*

Terminating the cable

The cores in flexible cables are all insulated and twisted throughout the length of the cable and so care needs to be taken not to damage the insulation or the fine strands of conductor. The cpc cannot be used as a guide and due to the twist, pulling the cpc is not a practical option.

Stripping back a flexible cable is carried out using a solid knife to score around the circumference of the cable taking care not to cut right through the cable sheath and so damage the core insulation.

Figure 4.8 *Do **NOT** cut right through – only score the sheath*

Bending the cable on the score mark will now cause the sheath to separate.

The part that is to be removed can now be pulled over the cores and completely removed as a tube. The length of sheath that can be removed in this way depends on the csa and number of cores in the cable. For smaller cables a length up to 100mm is quite achievable, whilst

Figure 4.9 *Flex to separate the remaining sheathing*

for larger cables 50mm of sheathing is more reasonable. Where longer lengths are required the outer sheath will have to be removed in stages.

The insulation can be removed from the cores in the same way as for solid wiring cables.

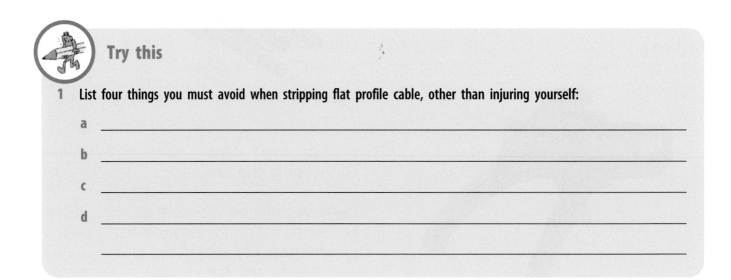

Try this

1 List four things you must avoid when stripping flat profile cable, other than injuring yourself:

a _____

b _____

c _____

d _____

_____

Part 2 Termination techniques for mineral insulated cable

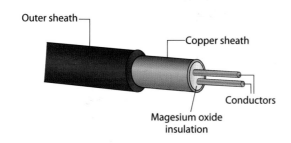

Figure 4.10 *Sheathed MIMS cable with copper cores and sheath*

Terminating the cable

The outer sheath, which may be either thermoplastic or thermosetting, is an option and therefore not included on all cables. As discussed in the previous chapter the outer sheath may be necessary for the environmental conditions in which the cable is to be installed and operating. The cable gland would also be covered with a shroud once the termination is complete, however this must be fitted before the cable is terminated so it can be slid into position later.

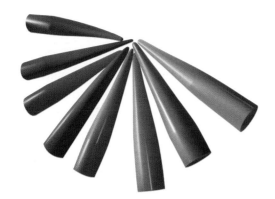

Figure 4.11 *MIMS shrouds*

The shrouds are sized so that one size will fit a number of different cable sizes and are supplied with one end closed. This closed end must be trimmed to suit the outer diameter of the cable which should be a good tight fit as it is to

continue the protection for the exposed metal sheath and the gland.

Where an outer sheath is fitted this must first be cut back to give enough exposed cable to allow for termination. The sheath should not be removed beyond the length which will be contained within the shroud once the termination is complete.

The termination of the cable requires some measurements to be taken to ensure the finished termination is correctly positioned.

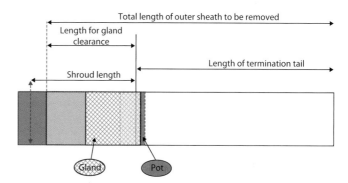

Figure 4.12 *Measurements for the termination*

> **Remember**
> It is essential to terminate this type of cable in a dry atmosphere as the insulation will absorb moisture.

Once the outer sheath has been removed the exposed metal sheath has to be removed. There are a number of special stripping tools available for this purpose. Each type of tool works on a principle similar to a tin opener. A blade cuts into the metal sheath which is then peeled off as the stripping tool is rotated.

When the correct length of stripped cable has been reached a pair of pliers can be used to

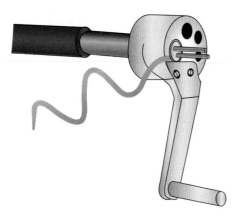

Figure 4.13 *MIMS cable stripping tool*

stop the tool moving down the sheath. This will terminate the process and leave a clean cut on the cable sheath. Any insulation left attached to the conductors should be tapped off at this stage.

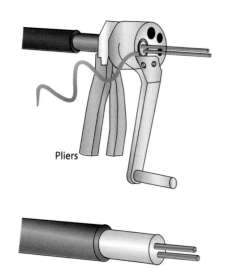

Figure 4.14 *Outer metal sheath removed*

The complete termination and seal components and sequence is shown in Figure 4.15.

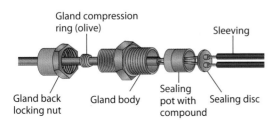

Figure 4.15 *The complete termination (without an outer sheath)*

If a complete termination is being used, then the termination gland should be slid onto the cable

in the correct sequence as shown in Figure 4.15 with the shroud fitted first.

The sealing pot shown in Figure 4.16 is screwed onto the sheath of the stripped cable. As the gland and pot are sized for the particular csa and number of cores in each cable the correct size must be used for the cable being terminated.

Figure 4.16 *Sealing pot*

A pot wrench may be used for this purpose which uses the body of the MIMS gland to align the pot squarely onto the sheath.

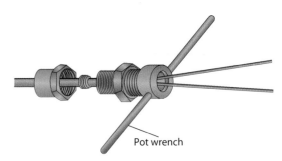

Figure 4.17 *Pot wrench uses the gland body to align the pot*

The sealing pot must now be filled with sealing compound. Before the compound is put into the pot the conductors need to be cleaned with a clean lint free cloth to remove any traces of the magnesium oxide insulation. The pot should also be tapped gently to remove any insulation residue dislodged in the fitting process. Care must be taken at this stage to eliminate any air pockets inside the sealing pot. This is achieved by filling the pot from one side only which forces the compound through the pot and expels the air at the same time.

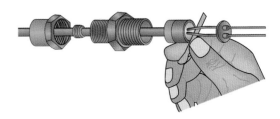

Figure 4.18 *Filling the sealing pot*

This must now be sealed with the plastic disc and crimped in place so that it cannot be pulled off and this is done using a crimping tool which is generally either a screw type or a set of crimp pliers.

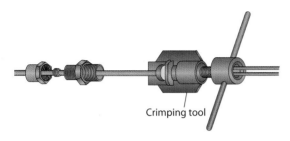

Figure 4.19 *Crimping tools for MIMS*

The conductors now require an insulated sleeving for their full length up to where they will be finally connected to a terminal. The sleeving is neoprene and supplied in hanks which can be cut to length and is normally coloured black.

Figure 4.20 *Finished termination*

The termination must be tested to confirm that the termination has been completed correctly and there are no connections between conductors or between conductors and the metal sheath.

> **Note**
> To be on the safe side, it is advisable to test the MIMS cable before crimping the sealing disc onto the pot. This test will also be repeated once the crimping is completed.

> **Remember**
> The other end of the cable will need to be stripped back to ensure there are no faults introduced at that end of the cable. As the cable is hygroscopic even a clean cut end will absorb moisture and appear as a fault when testing.

As the conductors are not marked it is necessary to test the conductors to identify them to ensure correct polarity and/or phase sequence when connected to equipment.

Connection to accessories

In standard accessory boxes where there is a 20mm diameter hole, the complete termination can be used and tightened with locknuts or a coupler and male bush. The method used may depend on the size of the accessory using the box.

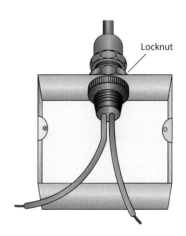

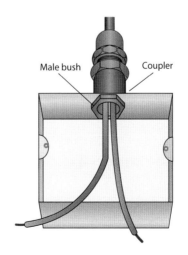

Figure 4.21 *Termination options for MIMS cable to accessory boxes and equipment*

Task

Using manufacturers' information from catalogues or online, list the materials necessary to terminate a 2L2.5 MIMS cable into a:

1 Surface mounted metal clad isolator.

_____

_____

_____

_____

_____

_____

2 Stop end conduit box for a luminaire.

_____

_____

_____

_____

_____

_____

_____

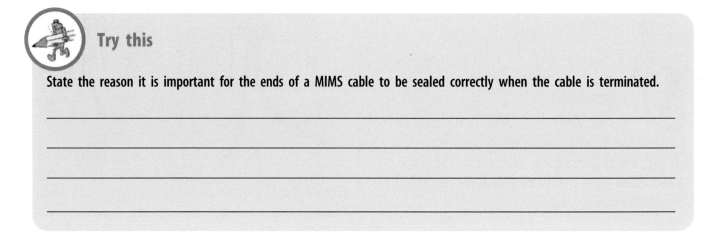

Try this

State the reason it is important for the ends of a MIMS cable to be sealed correctly when the cable is terminated.

_____

_____

_____

_____

Part 3 Termination techniques for steel and aluminium armoured cables

Armoured cables are widely used for industrial installations, distribution cables and underground cabling. They are generally available in the following formats:

- Steel wire armoured (SWA)
- Aluminium wire or tape armoured (AWA and ATA)
- Paper insulated, lead sheath cables (PILC).

PILC cables were widely used in the public distribution system for many years but their installation is now only used in special circumstances.

The armoured cables are available with thermoplastic (pvc) and thermosetting (cross-linked polyethylene or XLPE) insulation. The termination process is the same for both types of insulation.

The steel and aluminium armouring on these cables protect the live conductors from mechanical damage and may also be used as a cpc. However, the suitability depends upon the csa of the armour and this will depend upon the csa and number of cores of the cable conductors. Where the armour is not used as the circuit protective conductor (cpc) it forms part of the electrical installation and will need to be connected to the main earthing terminal. This may be done using either the cpcs or a separate conductor to earth the armour.

As the large sizes of these cables are very rigid the termination often has to be made within a particular tolerance as slack cable cannot be pushed back and there is no extra to spare. This means that the cable must be cut, stripped and terminated accurately.

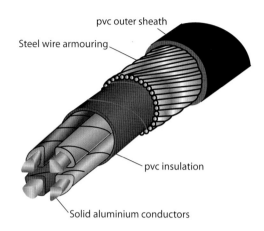

pvc outer sheath

Steel wire armouring

pvc insulation

Solid aluminium conductors

Figure 4.22 *Armoured cable*

Terminating the cable requires a cable gland and whilst there are several different types of armoured cable terminations they all have some similar features.

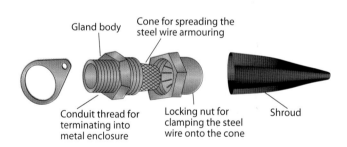

Gland body

Cone for spreading the steel wire armouring

Conduit thread for terminating into metal enclosure

Locking nut for clamping the steel wire onto the cone

Shroud

Figure 4.23 *Armoured cable gland*

To terminate an armoured cable within a set distance the gland body needs first to be fitted into the enclosure in the position it will be when the termination is complete. The gland locking nut, and the shroud if one is used, need to be slipped over the cable at this stage. These can be taped back out of the way for they will not be required yet.

Hold the cable where it will be installed and mark off the outer sheath at a point above the base of the cone, one thickness of the diameter of a steel wire, as shown in Figure 4.24.

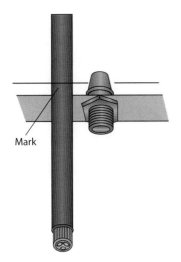

Figure 4.24 *Marking the outer sheath for stripping*

Cut a single line around the sheath at this point using a junior hacksaw, taking care to ensure that this cut is made only halfway through the steel wire armour. Alternatively, a proprietary

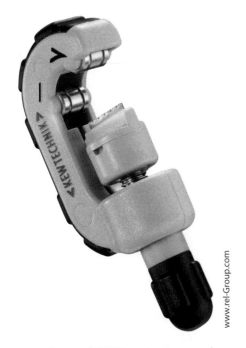

Figure 4.25 *Typical SWA stripping tool*

armoured cable stripper can be used. This is rather like a pipe cutter with a serrated blade. The depth of cut is controlled and so the risk of cutting through the armour is eliminated.

The outer sheath is then removed from this cut line to the end of the cable. For smaller cables this may simply be pulled off, for larger cables the sheath will need to be stripped using a knife. With the outer sheath removed the armours can be taken off at the cut. This is best achieved by bending the wires back and forth until they break. Care should be taken not to distort the steel wires within the sheath.

Once the wires have been cut, offer the gland body, the tapered section, against the remaining outer sheath and mark the sheath ready for removing the remaining section. This needs to be cut through and removed using a knife, taking care not to damage the conductors or insulation.

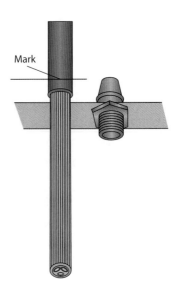

Figure 4.26 *Marking the remaining sheath*

The cable now has to be put through the gland body into its completed position. The steel wire strands have to be spread out over the cone making sure none are crossing over others.

On smaller cables this may be done by rotating the inner sheath and cores in the direction of the armour twist to open them enough to allow the tapered section of the gland to slide underneath the armour.

When everything is in place the gland locking nut can be moved into place and tightened up.

The inner sleeve can now be removed from the cable cores. This may be removed in the same way as the outer sheath of a flexible cable discussed earlier in this chapter. Care must be taken not to damage the insulation around the conductors.

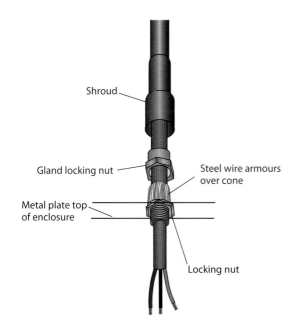

Figure 4.27 *Armoured cable terminations*

Earthing metallic cable sheaths and/or armour

Where metallic sheathed and armoured cables are terminated, the continuity of the earth for the sheath and/or armour must be maintained. The termination of a gland to a metallic enclosure is not in itself sufficient to do this and a reliable connection must be made between the metallic sheath or armour and the earth terminal

of the equipment or enclosure. This may be achieved in one of two ways:

● Earth continuity ring, often referred to as a 'banjo', which is installed below inside the enclosure between the gland lock ring and the enclosure.

TLC direct

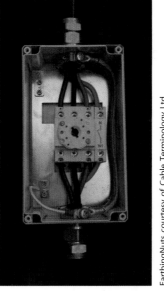

EarthingNuts courtesy of Cable Terminology Ltd
www.earthingnuts.co.uk

Figure 4.28 *Earthing connections for the armour or sheath*

A bolt is installed through the hole in the ring and a suitable protective conductor is connected between this bolt on the continuity ring and the earthing terminal.

- An earthing lock nut has a terminal contained within the locknut and once tightened a conductor can be installed between the locknut terminal and the earthing terminal of the enclosure. This has the advantage that additional holes and bolts do not need to be installed through the enclosure.

The correct use of such earthing arrangements is essential where earth continuity is to be maintained through an insulating enclosure, see Figure 4.28.

Paper insulated lead sheathed cables

PILC cables come in a number of construction formats relating to the method of protection provided over the lead sheath. The main types are steel tape armoured (PILCSTA) and steel wire armoured (PILCSWA). The termination of these cables is similar to that for other armoured

cables but requires some particular additional skills. Stripping the lead, removing the paper insulation and sealing the termination is a specialist activity and not covered in this study book.

Metal braided flexible cables

The termination of a braided flexible cable is similar to that for a SWA cable and the glands used for the termination are very similar. There are installations where the braid is used for mechanical protection and/or screening. The braid does constitute an exposed conductive part and should be treated as such and suitably earthed. However, there may be occasions where the braid is used for screening purposes where it is connected separately and therefore requires isolation at the equipment.

Metallic and insulating glands are therefore available for this cable and a typical metallic gland is shown in Figure 4.30.

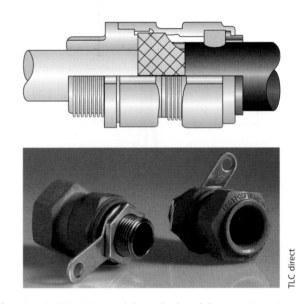

Figure 4.30 *Typical braided cable termination and gland*

The termination process is the same as that used for the armoured cables as described earlier in this chapter.

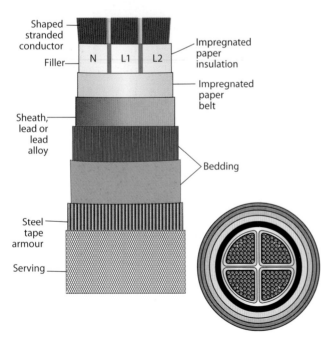

Figure 4.29 *PILCSWA cable construction*

Task

Using manufacturers' information from catalogues or online, list the type of gland required to terminate each of the following, together with the manufacturer's catalogue number.

1 A three core, 16mm^2 thermoplastic steel wire armoured cable.

_____

_____

2 A four core 25mm^2 aluminium armoured cable.

_____

_____

3 A three core 2.5mm^2 steel braided flexible cable.

_____

_____

Task

Using manufacturers' information from catalogues or online, sketch the components for an aluminium armoured cable termination labelling each part.

Try this

Explain how the earth continuity of the steel wire armoured cable will be achieved when terminated at a metal clad distribution board.

_____

_____

_____

_____

Part 4 Termination techniques for braided flexible cables, fire resistant cables and coaxial cables

There are flexible cables which have a braided finish which is not metallic but still provides additional protection to the outer sleeving of the cable. The braid is generally hard wearing and abrasion resistant and will need to be secured to prevent fraying and the possibility of contact with any live terminals. There are two basic methods of achieving this which are:

● Heat shrink sleeving
● Neoprene sleeving.

Heat shrink sleeving is the simplest method to neatly terminate the braiding. The sleeving reduces in size with the application of heat, generally to around 50 per cent of its initial size. The sleeving size is to be determined by the required finished diameter based on the diameter of the outer sheath of the cable. The sleeving is cut to a suitable length and slid over the braiding so that it extends onto the outer sheath of the cable.

Alpha Wire FIT® Heat-Shrink Tubing

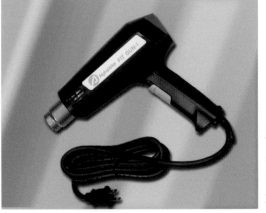

Alpha Wire Heat Gun

Figure 4.31 _Heat shrink sleeving and heat gun_

Heat is then applied using a heat gun and the sleeving shrinks to contain the braiding and grip securely to the outer sheath of the cable. This provides a neat and water resistant finish. A heat gun is necessary for this type of termination but the finished result is far superior.

Neoprene sleeving is often used for braided appliance connection cords. It consists of a neoprene sleeve which is stretched using a proprietary tool to allow it to be slipped over the braiding in the same fashion as the heat shrink sleeving. The tool pressure is then released and the tool withdrawn leaving the braid secured.

A stuffing gland may then be used to terminate the flexible cable into the enclosure once the braid has been terminated.

Figure 4.32 *Typical stuffing gland*

Fire resistant cables

These are often referred to as FP and PX cables and consist of an outer sheath over a layer of aluminium foil. The cores of the cables are similar to other sheathed and insulated cables consisting of insulated live conductors and a bare cpc.

The insulation retains its electrical integrity under fire conditions, however it does not retain its mechanical strength. The aluminium sheath

provides the mechanical protection and support under fire conditions. The tinned copper cpc is in contact with the aluminium sheath throughout the length and so the aluminium sheath does not have to be earthed separately.

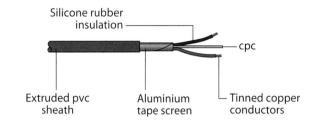

Figure 4.33 *Aluminium foil sheathed cable*

The cable is stripped by scoring around the outer sheath with a knife and then flexing the cable to and fro until the aluminium breaks off. The tube of aluminium and outer sheath can then be pulled off leaving the inner cores.

Where these cables have to be terminated into conduit accessory boxes and enclosures, a gland similar to the stuffing gland is used. This consists of a main body and locking bush which, when tightened, compresses a rubber 'O' ring onto the cable.

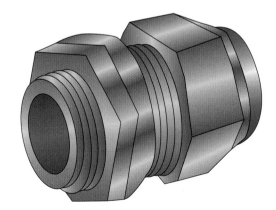

Figure 4.34 *Fire resistant cable securing gland*

Data cables

The majority of data cables use insulation displacement (ID) terminations which mean that there is no stripping back of insulation for the termination of the conductors. Where the cables have an outer sheath this is removed using either:

● The same process as for a multicore flexible cable, or
● Using a proprietary cable stripper.

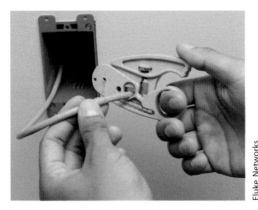

Fluke Networks

Figure 4.35 *Typical cable stripper*

Terminations into enclosures and equipment may then be made using a stuffing gland to ensure the cable is suitably supported and protected and the integrity of the enclosure is maintained.

Courtesy of Trend Control Systems Ltd

Figure 4.36 *Typical twin twisted pair*

Coaxial cables

Coaxial sockets and plugs and other similar connections are used on screened cables where the screening is connected to the outside metal and the core to the centre pin.

Figure 4.37 *Typical coaxial socket and plug*

A coaxial cable has inner and outer cylindrical conductors and tools are available for stripping the cable ready for connecting to the socket or plug. To connect the socket or plug first unscrew the connector and slide the screw cap over the cable. Then strip away about 23mm of the outer sheath and gather the strands of the copper screening, winding them back around the outer insulation. About half of the inner insulation should now be stripped and the screening wire should not touch the inner copper wire. Fit the claw, which clamps the outer screen, over the inner sheath and sit it around the screening. Push the inner copper wire through the hole in the remaining half of the plug and, when the screw cap is tightened, the outer screen should be trapped by the claws inside the plug. Any protruding wire should be cut off flush with the plug.

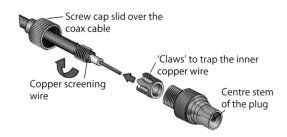

Labgear

Figure 4.38 *Typical coaxial cable stripping tool*

Figure 4.39 *Typical coaxial cable and plug connections*

Try this

Sketch the construction of the following cable types with the appropriate termination, labelling all the component parts:

1   Fire resistant cable.

2   Coaxial cable.

Part 5 Termination techniques for fibre optic cables

Fibre optic cables have a complete termination process which involves both a scrupulously clean environment and equipment to ensure a suitable and effective connection. This process does vary from one manufacturer to another and the manufacturer's instructions should always be followed closely.

There are a number of different terminations for fibre optic cables and the most popular are:

Straight tip (ST): ST connectors have a key which prevents rotation of the ceramic ferrule and a bayonet lock. The single tab must be properly aligned with a slot on the socket before insertion and then the bayonet lock is engaged by pushing and twisting. This locking at the end of travel maintains a spring-loaded section which keeps the pressure on the optical core junction.

Figure 4.42 *Typical LC connector*

Figure 4.40 *Typical ST connector*

Square connector (SC): sometimes referred to as subscriber connector have a snap-in with a simple push–pull design which reduces the chance of fibre end face contact damage during connection.

Figure 4.41 *Typical SC connector*

Little connector (LC): these are replacing the SC connectors in corporate networking environments due to their smaller size which uses a 1.25mm ferrule which is half the size of the ST connector and they are easily terminated.

There are other varieties of connector but they all operate on a similar principle and the termination of the fibre optic is much the same in each case.

We will consider the basic requirements for a termination here but as stated earlier the manufacturer's instructions for the particular termination must be closely followed.

Basic equipment

You will need a suitable termination kit which should contain:

- Fibre optic connectors
- Carbide scribe
- Kevlar shears
- Lint free cloth
- Pure alcohol solution for surface cleaning
- Polishing surface (usually a glass plate and a rubber mat)
- Polishing disc
- Polishing papers (1, 3 and 5 microns)
- Syringe for the epoxy adhesive
- Epoxy adhesive
- Adhesive activator
- Fibre optic crimp tool
- Fibre optic stripper
- Fibre optic microscope.

Fibre Optic Termination Kit, courtesy of Greenlee Textron Inc.

Figure 4.43 *Fibre optic termination kit*

This equipment is for a cold cure adhesive, some manufacturer's equipment requires a connector heater as part of a hot curing process for the termination.

As we can see, a knife, a pair of pliers and a pair of sidecutters are not going to be enough for this type of conductor termination and there are some specialist tools and equipment needed.

The basic steps for the termination are:

- Place the boot and crimp on the cable small end first
- Strip the outer sheath from the cable using a suitable cable stripper
- Twist the Kevlar fibre, sight and trim to length with Kevlar shears
- Strip the fibre buffer, using the stripper, to leave the required length in place (strip in short sections to prevent breaking the fibre)
- Clean the fibre with a lint free cloth to remove any residual coating material and lay to one side on the lint free cloth
- Place the required amount of adhesive into the syringe

- Insert the plunger, turn upright and expel the air from the syringe, taking care to catch any spillage from the tip in a clean cloth
- Place the syringe needle in the rear of the ferrule and squeeze the plunger until a small bead of adhesive appears at the ferrule tip, making sure no adhesive goes into the gap between the inner and outer sleeves of the connector
- Apply the activator to the fibre; this is usually an aerosol spray application, taking care to contain any overspray
- Quickly and carefully push the fibre into the ferrule until the buffer reaches the back of the connector
- Hold the fibre in position to allow the adhesive to cure
- Score the fibre (don't cut) at the point where it emerges from the ferrule
- Pull the fibre away from the ferrule (in line with the ferrule) which will detach the scored fibre
- Place the fibre in a sharps container
- Slide the crimp collar back over the cable and the Kevlar until it is against the shoulder of the connector
- The crimp collar is then crimped onto the cable in two stages, the large front section first and then the smaller rear section
- Slide the boot up the cable and over the whole crimping collar
- Gently polish the end of the fibre with the coarse polish cloth to remove the rough end
- Place the glass plate and rubber mat on a firm surface
- Place the medium polishing paper on the rubber mat (a drop of de-ionized water on the mat helps secure it) and place a drop of de-ionized water on the surface of the polishing paper
- Clean the surface of the polishing guide (this is part of the termination pack) and fit the connector ferrule into the hole in the guide

- Place the ferrule tip on the polishing paper and using light pressure move in a figure of eight for around 25 rotations to polish the end of the fibre
- Remove the ferrule from the guide and clean the tip of the fibre and the guide using a lint free cloth moistened with the alcohol
- Check the end of the fibre using a fibre optic microscope (cross between a microscope and a telescope to look at)
- Fine scratches can be removed with further polishing
- If there are chips or cracks these cannot be removed and the terminal will need to be cut off and a new termination fitted

- If all is well repeat the polishing process using the fine polishing paper
- Check the end of the fibre and this should be scratch free
- Fit the protective cap over the fibre until it is connected into the equipment.

As we can see the termination process is quite involved and requires a level of skill. It is advisable to practice the termination process until you are familiar with it and able to produce acceptable terminations before carrying out terminations on site.

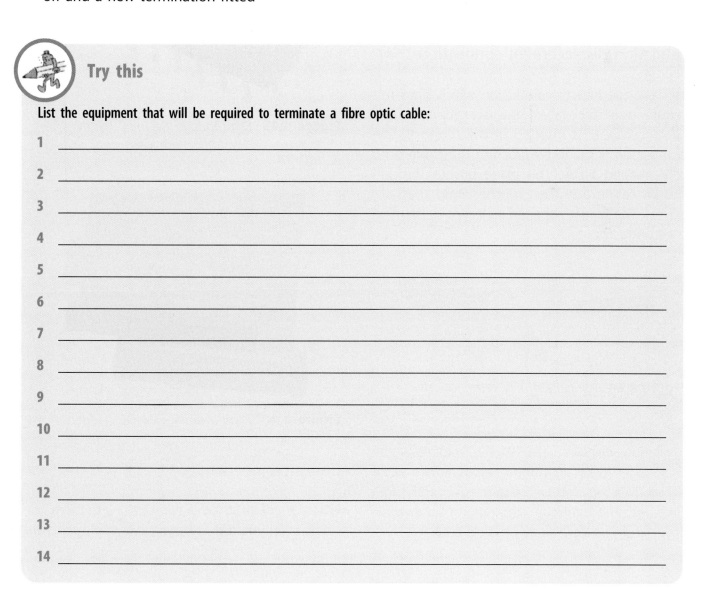

Try this

List the equipment that will be required to terminate a fibre optic cable:

1 _____

2 _____

3 _____

4 _____

5 _____

6 _____

7 _____

8 _____

9 _____

10 _____

11 _____

12 _____

13 _____

14 _____

Part 6 Terminals and terminations

Terminations

There are specific terminations which are used for the connection of conductors within the electrical installation industry. We have considered the termination of fibre optic cables in some depth as they are a complete and specialized process of their own. In a similar vein are the ID terminations which are largely used in data cable terminations within the IT industry and in telecommunications.

Insulation displacement connection

This is a method of connection which does not require the removal of the insulation from a cable before it is inserted in the terminal. The connection is made by pressing the cable onto a 'V' shaped blade. This blade pushes through the insulation and then touches onto the conductor. The blade is shaped so as not to damage the conductor.

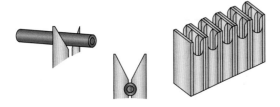

Figure 4.44 *Principle of IDC terminations*

Insulation Displacement (IDC) terminations are used extensively for ribbon cables where all of the conductors can be connected at the same time with the minimum separation of individual cores and in a very small physical area.

To ensure good electrical and mechanical connections special tools are used for inserting the cables into the connectors, the most basic being a handheld single core tool. This is often used for terminating telephone cables.

Figure 4.45 *Dual-in-line (DIL) IDC header*

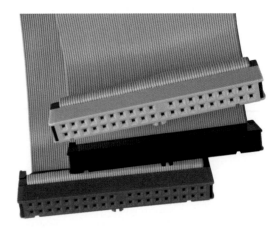

Figure 4.46 *Dual-in-line (DIL) IDC header*

Crimp connections

Where there are a lot of connections to be made, such as terminating a ribbon cable, hand tools are available for crimping these multi-way connectors. One type holds the connector and the cable in position in a bench vice. When the jaws of the vice are closed the outer insulation is pierced and the connection is made to the inner conductors. A strain relief clip is usually fitted to complete the termination. Any excess ribbon cable should be trimmed off neatly using cutters or a sharp knife.

Hand crimped connections are carried out using a purpose-made crimping tool. It connects each individual wire to a multi-way connector contact which is inserted into a connector housing.

Often there is a separate tool available for removing the contact from the connector body and many manufacturers have specific tools for their products.

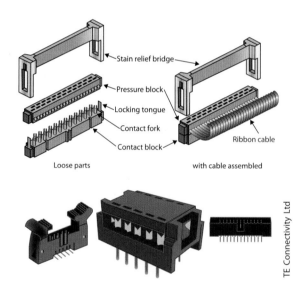

Figure 4.47 *Crimp ribbon connectors*

TE Connectivity Ltd

Stain relief bridge
Pressure block
Locking tongue
Contact fork
Contact block
Ribbon cable
Loose parts
with cable assembled

Terminals and connections

Connections fall into two main categories, those which are:

● Readily separable by unplugging or disconnecting (plugs and connectors)
● Mechanically fixed such as soldered and wire wrapped.

Many of the data cables are connected by means of plugs and socket terminals despite the actual terminations being mechanically fixed by crimping, adhesive and IDC. It is fair to say that many of the internal connections within the equipment are soldered.

The most common termination methods used in electrical installation work are those which require a mechanical fastening by a form of screw or bolt. These may be coupled with other methods of terminating the cable in readiness for the final termination.

The main methods of connection, together with their advantages and disadvantages are the next considerations.

Tunnel terminals

These are probably the most common and have been used in electrical equipment such as switches, sockets, ceiling roses and luminaires for over 100 years. The key point to remember when terminating at tunnel terminals is that the conductor should fill as much of the terminal as possible. Generally, terminals are able to accommodate the largest conductor likely to be installed and often more than one of these conductors in each terminal.

For example a BS 1363, 13A socket will accommodate a 4mm^2 conductor and up to three 2.5mm^2 conductors quite comfortably. When a single 2.5mm^2 conductor is to be terminated into a socket it is important to make sure the cable is securely held by the terminal screw and a good electrical connection is made. This is generally achieved by using pliers to double the conductor back on itself as shown in Figure 4.48. This effectively increases the conductor area and provides a good mechanical and electrical connection. For higher current applications, tunnel terminals may have more than one screw per terminal to ensure a reliable connection.

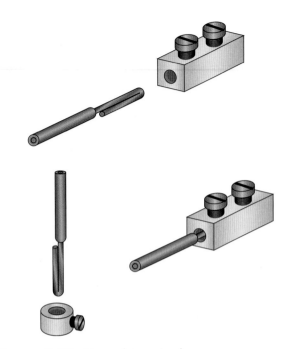

Figure 4.48 *Tunnel terminals*

One of the possible problems with this type of terminal is that smaller conductors can be held in the terminal but fail to make a good electrical connection as they are not below the terminal screw as shown in Figure 4.49.

Figure 4.49 *Conductor held but not electrically sound*

Pressure plate terminals

These are another variation of the tunnel terminal. The pressure plate terminal is shown in Figure 4.50. The terminal is opened and the conductor is placed between the back of the

terminal and the plate. When the screw is tightened the plate is pushed forward and clamps the conductor against the back of the terminal. These terminals are common on the outgoing connections of circuit breakers, residual current devices (RCDs) and the incoming tails of consumer units.

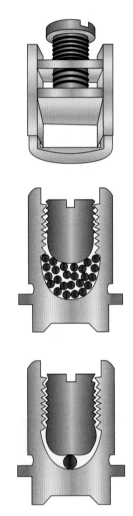

Figure 4.50 *Pressure plate and screw pinch terminals*

Screw pinch terminals

These are another variation of the tunnel terminal only for this type the terminal is a threaded open U shape. The screw can be fully removed and the conductors are laid in the terminal and the screw replaced and tightened. This type of terminal is common in screw terminal joint boxes which allows straight-through conductors to be connected without cutting them. The insulation is simply removed for the length of the terminal and this is shown in Figure 4.51.

Wrap around terminations

These terminals rely upon the conductor being wrapped around a screw or bolt. These terminals were commonly used in equipment and generally for termination of protective conductors. The conductor should be wound onto the bolt or screw in the direction in which the screw or nut is turned to tighten it. This will pull the conductor into the termination and so aid the tightness of the finished terminal. If the conductor is wound in the opposite direction the conductor tends to pull out of the termination as it is tightened.

A variation involves the use of a crown washer and plain washer. The conductor is formed into a loop and placed inside the crown washer and the plain washer placed on the top. The crown washer is then crimped down onto the plain washer firmly holding the conductor between the two washers as shown in Figure 4.52.

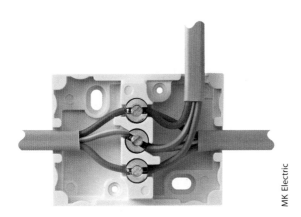

MK Electric

Figure 4.51 *Typical three terminal joint box*

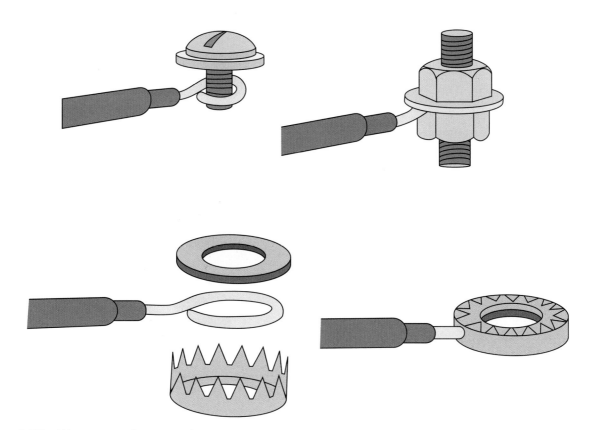

Figure 4.52 *Wraparound connections*

Crimp and solder lugs

These provide a means of connecting conductors to terminals such as the screw and bolt considered in the wraparound terminals. Crimp terminals provide a dry and relatively quick method of terminating conductors. As we discussed earlier they may be used to permit termination of dissimilar metals or to enable a suitable connection to be made. A large aluminium conductor to be connected to a bolted terminal in a fused switch could not be made using the wraparound method and a crimp lug connection is a quick and effective method of achieving this.

The crimp lug must be selected to suit the conductor csa and shape in order for a good connection to be made. The crimp tool must also have the appropriate die to ensure a good connection. For smaller csa cables a ratchet hand crimp tool may be used. For larger csa conductors a hydraulic crimper is used which has interchangeable dies to suit different csa and conductor shapes.

Solder lugs

These have largely been replaced by crimp lugs as they perform similar functions but require the application of heat and the use of solder flux to complete the termination. They were mostly used on copper conductors and there are many still in service.

www.ultimatehandyman.co.uk

CEMBRE LTD - www.cembre.co.uk

Fastenright

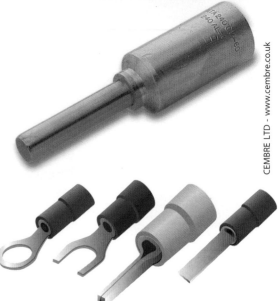

Figure 4.53 *Various types of crimp lug*

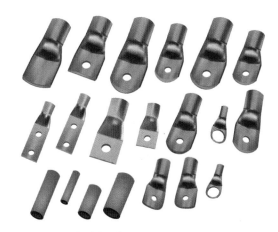

Figure 4.54 *Solder lugs*

Solder connections

Solder connections are used in many applications, particularly in the manufacture of electrical and electronic equipment. Terminations are made to printed circuit boards (PCBs) and in many cases the connection to the outside world is made using plug-in connections with the connector soldered to the circuit board. We will consider some of the methods used to make these soldered connections.

Connecting to printed circuit boards

A PCB is an insulated base with tracks of conducting material running in an intricate pattern over one or both sides of the board. There are also multi-layer boards with up to as many as 10 layers.

Where PCBs are made as a replaceable unit they often incorporate a plug-in arrangement on the board. One example of this is the edge connector. Here the PCB has been made so that the track goes out to the edge of the board. To ensure a good electrical contact the area that plugs into the socket is often gold-plated.

Task

Using manufacturers' information from catalogues or online, list one application for each of the following termination methods:

1   An IDC termination.

_____

2   A tunnel terminal.

_____

3   A wraparound connection.

_____

4   A pinch plate termination.

_____

5   A crimp lug.

_____

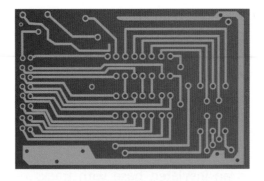

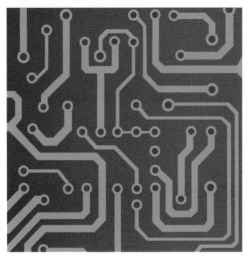

Figure 4.55 *Printed circuit boards*

Figure 4.56 *Edge connectors on a PCB*

PCB should always be handled with care. The manufacturing process of the copper tracks means that they may be very thin in places and can easily be damaged. Many of the components on the boards may be damaged by the minute voltages in the human body. The boards must be handled by their edges without touching the conductor tracks. It may be necessary to wear an earthed strap to stop the effects of static electricity from damaging the sensitive components.

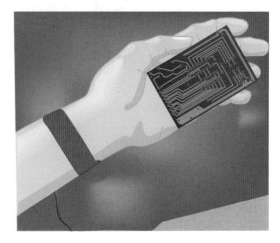

Figure 4.57 *Correct handling of PC boards*

Circuit components or their mounting bases are connected directly onto the board by soldering. Great care must also be exercised when replacing the components which have been mounted on a PCB. The copper tracks are very fine and can be fractured if the board is bent, flexed or overheated during the process.

Solder was traditionally an alloy of tin and lead used to bond metallic materials together. The solder used for electrical soldering was 60 per cent tin, 40 per cent lead which melts at 188°C. To ensure a good electrical and mechanical connection a resin flux is used to prevent oxidation of the metals.

However on 1st July 2006 the European Waste Electrical and Electronic Equipment Directive (WEEE) and the Restriction of Hazardous Substances (RoHS) came into effect. These prohibit intentional addition of lead to most consumer electronic goods manufactured in the EU. As a result, lead free solders were introduced and the move to a greener environment has promoted their use in most areas of industry. Lead free solders used commercially may contain tin, silver, copper, zinc, antimony and other metals.

The solder in general use is multicore 22 s.w.g. and has the flux run in cores through the length of the solder. There are other solders that exist for a variety of temperatures and purposes. For example, in response to the needs of a 'greener' environment there is now also halide free flux.

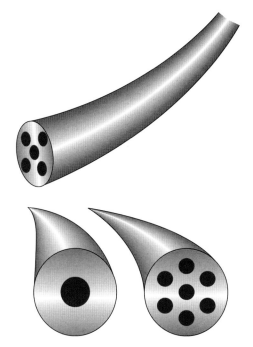

Figure 4.58 *Common varieties of multicore solder*

Soldering techniques

There are two stages to making a good solder connection. First, the separate parts of the joint must be prepared and second, they need to be put together and soldered. Most conductors will require tinning before they are to be soldered. This involves heating the conductor and applying a small amount of solder to cover the surface. This process also cleans any surface impurities away to ensure a good joint.

There are two types of conductors:

● Single strand
● Multi-stranded.

Single strand conductors should have any insulation removed to the required length and then the conductor is usually ready for tinning and connection.

Stranded cable will, in addition to removing the insulation, need the copper strands to be twisted together and tinned. To tin the wire the soldering iron and solder are applied to the end of the twisted wire and then worked up towards the insulation. Tinning stranded conductors is a skill that needs practice as the amount of solder and heat used is critical. A well tinned conductor should result in a bright shiny surface where all strands are joined together as one.

Part of the preparation of any joint is forming the conductors to the required shape. There are several methods of soldering conductors to solid terminations and we shall consider three of these.

Soldering to a pin

The wire should be bent to a hook shape using long nosed pliers and the hook placed over the pin and squeezed to form a tight fit.

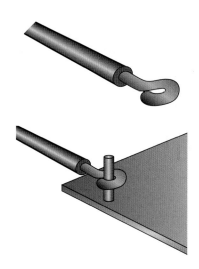

Figure 4.59 *Pin connection*

Soldering to tag strip

The conductor is threaded through the hole in the tag strip and then bent round the tag and any excess wire is cut off and the joint squeezed flat with pliers. With some components it may be more appropriate to bend the connecting wire 90° for ease of removal or replacement.

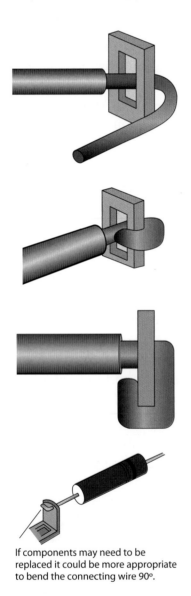

If components may need to be replaced it could be more appropriate to bend the connecting wire 90°.

Figure 4.60 *Tag connection*

Soldering to printed circuit board

If a component is to be fitted to a PCB the leads must be formed to the correct position before being inserted. This can be carried out using a pair of long nosed pliers or special lead

forming tools. The forming tool consists of two arms over which the component leads are bent. When a number of components all have to be prepared the forming tool ensures a consistent lead position on all components.

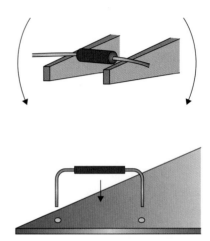

Figure 4.61 *Forming the component*

Once the component has been inserted into the board the leads should be bent to keep the component in place. The leads can now be cut back leaving enough to hold the component but not so much to short out other tracks on the PCB.

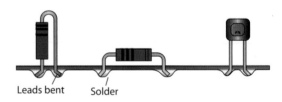

Leads bent   Solder

Figure 4.62 *Components soldered into position*

There are three main factors for a good soldered connection:

● Cleanliness
● Correct heat
● Correct flux (if a cored solder is used this should be ensured).

Cleanliness at the joint can be increased by tinning the surfaces first. That is to coat them in a thin layer of solder.

Remember

Aluminium conductors cannot generally be soldered in this way.

For good soldering joints

- The iron should be clean and fluxed
- When solder is applied to the iron it should appear bright and shiny if the iron is up to temperature
- The surfaces to be soldered should be clean and tinned
- The iron should be applied to the surfaces so not to be connected to the solder
- Solder should be seen to run over the surfaces when the joint is made
- The joint should be allowed to cool, not cooled by blowing or applying a damp cloth.

Precautions when soldering

- Never flick off solder from the iron as this may cause harm to persons, clothing or apparatus
- When soldering inside equipment it may be necessary to cover components
- Heat shunts may need to be applied on some heat sensitive components
- Always keep the soldering bit clean and tinned
- Use the soldering iron stand to prevent danger from burns or fire

- Use a proprietary cleaner for removing flux residues so that the joint can be readily inspected.

Common soldering faults

- Dry joints – these are generally caused by not using a high enough temperature or by moving the leads before the solder has set. All surfaces have to reach a temperature high enough to melt the solder. A dry joint often looks grey and dull not bright and shiny as it should be.
- High resistance joints – dry joints are often also high resistance joints. It is not always possible to detect a high resistance joint visually. Where the surfaces have not been cleaned and tinned correctly it is possible for crystallization to form between the surfaces, particularly if a solder is used without sufficient flux.
- Excess solder – can lead to connections being shorted out accidentally.
- Too much heat – can cause several problems including:
  - Damage to insulation
  - Damage to the PCB
  - Hardening of the wire making it brittle
  - Damage to components – components may be protected by using a shield or a heat shunt.

A heat shunt looks rather like a crocodile clip and is clipped on the lead of a component before the lead is soldered to the circuit board. This diverts much of the heat travelling up the lead from the soldering iron and keeps the component relatively cool.

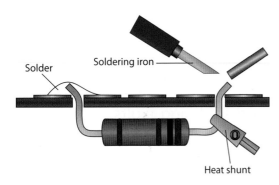

Figure 4.63 *The use of a heat shunt*

Desoldering techniques

Often when repairing electronic equipment it is necessary to remove a component from a PCB. Two methods of desoldering are commonly used. The simplest to use is the desoldering braid. The braid is placed on the joint and the soldering iron is placed on top to heat both the braid and the joint. When sufficiently heated the molten solder is wicked up into the braid.

The second method is to use a desoldering tool. This has a spring-loaded plunger that works like a bicycle pump in reverse and sucks hot solder from the joint when the plunger is released.

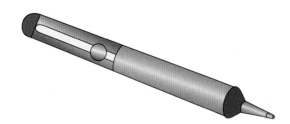

Figure 4.64 *Desoldering gun*

If excessive heat is applied when desoldering, the tracks can become detached from the board and broken. In addition to this, the components themselves can be damaged by heat.

Common desoldering problems

- If the desoldering tool is full of solder it does not operate smoothly
- When the joint is not hot enough solder remains on the joint. If this happens it is often better to re-solder the joint before desoldering again
- When the joint is too hot the PCB tracks may lift from the board or the components may be damaged
- When the desolder nozzle does not seal to the board little suction is achieved
- Some people leave the iron on the joint whilst applying the desolder nozzle – this can work with care but tends to damage the nozzle and the PCB as the desolder nozzle hammers the iron onto the PCB surface.

Having considered the most common methods of termination, Table 4.1 identifies the main advantages and disadvantages of the various types of termination and it is not intended to be a definitive list. Of course the most appropriate termination method must be selected for the location, terminal, conductor size and shape and the external influences to which it is to be subjected.

Remember

Terminations which are not electrically and mechanically sound may result in high resistance joints and/or corrosion and erosion of the materials. Good terminations are vital to ensure the good, effective and reliable connections throughout the life of the installation.

Try this

1 a List three important considerations which must be made to ensure a good soldered joint:

_____

_____

_____

b List three common faults in soldered joints:

_____

_____

_____

_____

_____

_____

_____

2 List two methods of desoldering:

_____

_____

_____

Table 4.1 *Main advantages and disadvantages of terminations*

Type of terminal	Advantages	Disadvantages
Tunnel terminal	Cheap to produce	Undoing the screw too far and the screw falls out, particularly where terminals are overhead, and can be easily lost and difficult to refit.
	Reliable	The terminal screw applies downward and rotary pressure on the conductor. This reduces the diameter of the conductor and weakens it so that if the conductor has to be disconnected re-termination may be necessary before reconnection.
	Easy to terminate	Smaller conductors can be held in the terminal but fail to make a good electrical connection as they are not below the terminal screw as shown in Figure 4.49.
	Conductors readily terminated using basic tools (screwdriver)	Over-tightening of the terminal screw can sever the conductor or shear the screw.
		Under-tightening of the terminal screw can result in overheating and arcing.
Pressure plate	Relatively cheap to produce	Larger conductors can become loosened in the terminal by the physical movement of the conductor. Some manufacturers fit two screws to the terminals for larger conductors.
	Reliable	Over-tightening of the terminal screw can damage the thread in the terminal.
	Easy to terminate	Under-tightening of the terminal screw can result in overheating and arcing.
	Conductors may be readily connected and disconnected using basic tools (screwdriver)	
	Minimal damage to the conductors	

Table 4.1 *(Continued)*

Type of terminal	Advantages	Disadvantages
Screw pinch terminals	Relatively cheap to produce	The terminal screw applies downward and rotary pressure on the conductor. This reduces the diameter of the conductor and weakens it so that if the conductor has to be disconnected re-termination may be necessary before reconnection.
	Reliable	Over-tightening of the terminal screw can sever the conductor or shear the screw.
	Easy to terminate	Under-tightening of the terminal screw can result in overheating and arcing.
	Conductors may be readily connected and disconnected using basic tools (screwdriver)	
	Through conductors need not be cut to terminate	
Wraparound connections	Cheap to produce	More involved preparation for conductor termination.
	Reliable	Relies on correct construction and fitting of the conductor.
	Conductors may be readily connected and disconnected using basic tools (screwdriver, pliers, spanner)	Under-tightening of the terminal screw can result in overheating and arcing.
Crimp lug	Quick and convenient to install	Requires special tools to fit the crimp.
	Secure termination	Tools for larger sizes are expensive and often hired.
	Can provide termination for dissimilar materials	
Solder lug	Good electrical connection	Requires the use of a heat source, usually a gas torch, and involves molten metals.
	Robust joint once set	Flux presents a hazard.
Soldered connections	Good electrical connection once set	Requires the use of heat, generally a soldering iron
	Mechanical strength	Molten metals and flux
	Small area of joint	Damage to components
		Damage to insulation

Try this

Name the connections or tools shown below

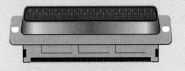

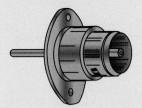

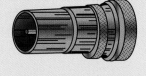

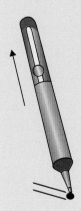

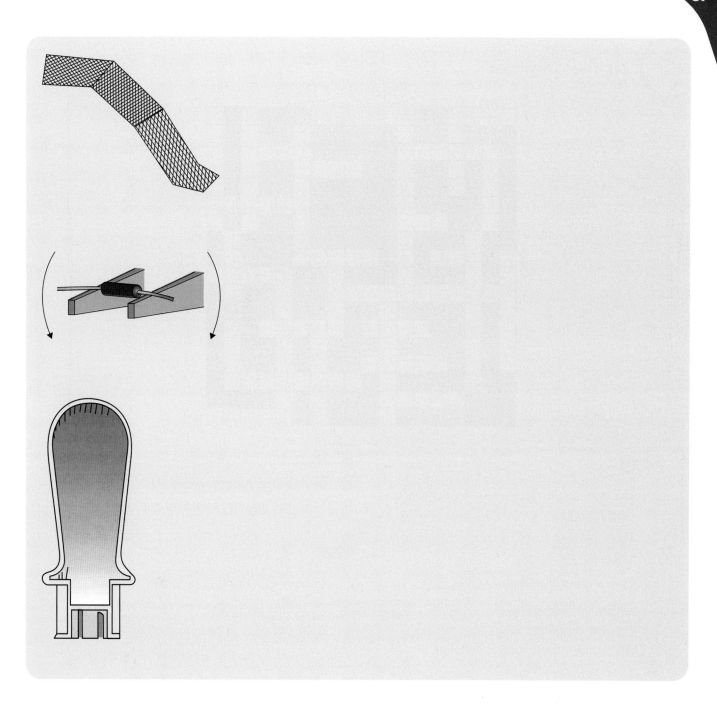

Try this: Crossword

Across

1   This type of cable has the best fire performance properties (4)

3   These are used to cut the Kevlar in fibre optic cables (6)

7   Used to cover the glands of cables (6)

8   Ribbon cables carry this (4)

9   Prevents unsheathed cables from damage where they enter a metallic enclosure (4)

10 Green and yellow or where your arms go? (6)

11 Used to hold the sealing compound in place (3)

15 Slang for an ear or a form of conductor termination (3)

16 The glass in optic cables is in this form (5)

18 A screw ... is a type of terminal (5)

19 Used to squeeze material into a small space (11)

21 Without insulation or sleeving (4)

22 Behind the bar in the pub or part of a data cable (5)

25 The S in SWA (5)

26 Pliable (8)

27 Used to strip cable sheathing (5)

28 Terminates the cable referred to in 25 across (5)

Down

2   Common component in removable terminals (5)

3   Electricians use these to cut cables (11)

4   Plaits and types of cable protection (7)

5   A type of connection applied by 19 across (7)

6   The fixing medium used in optic cables (8)

7   Used to prevent components overheating when soldering (5)

12 Lightweight armour? (9)

13 The copper covered version of 1 across (4)

14 Can be used to hold the gland securely (7)

17 Used with 14 down to provide earth continuity (5)

18 What you have to do to the end of an optic termination (9)

20 Shown in the cinema or preventing interference (8)

23 Used to twist conductors together (6)

24 At the heart of a cable (4)

Congratulations, you have now finished this chapter so complete the self assessment questions before moving on to the next chapter. As this chapter covers a number of topics there are ten questions in this self assessment.

SELF ASSESSMENT

Circle the correct answers.

1 Before the outer sheath of a flexible multicore cable can be removed it must be:

a. Scored with a sharp knife where the sheath is to be removed

b. Cut pushing a knife blade against one of the conductors

c. Snipped and pulled back using a pair of side cutters

d. Heated up to make it more pliable for stripping

2 An MIMS cable must be sealed when terminated to:

a. Prevent the conductors moving during termination

b. Provide a bedding for the insulation

c. Prevent the ingress of moisture

d. Keep the insulation in place

3 When cutting the armour during the termination of a SWA cable, the cut should go through the armour:

a. One-quarter of the way

b. Halfway

c. Three-quarters of the way

d. Completely

4 Fire resistant cable has the conductors contained within a tube of:

a. Plastic

b. Copper

c. Steel

d. Aluminium

5 IDC terminations are made using:

a. An insertion tool

b. A screwdriver

c. A knife

d. A vice

6 The type of terminal found in a standard through joint box is a:

a. Screw pinch

b. Pressure plate

c. Wraparound

d. Tunnel

7 A crimp lug must be selected to suit the conductor csa and:

a. Temperature

b. Diameter

c. Length

d. Shape

8 Insulation should be removed from a conductor entering a tunnel terminal so that it:

a. Just enters the terminal

b. Finishes at the terminal edge

c. Is trapped by the terminal screw

d. Finishes 3mm short of the terminal

9 Connections to a PCB which may be replaced are often made using:

a. Soldered joints

b. Edge connectors

c. Pin terminals

d. Crimp lugs

10 A small component is to be replaced on a PCB. To protect the component during the soldering process will require a:

a. Soldering braid

b. Anti-static bond

c. Heat shunt

d. Heat mat

5

Health and safety

RECAP

Before you start work on this chapter, complete the exercise below to ensure that you remember what you learned earlier.

● When stripping multicore cables it is important not to damage the _____ or the _____.

● When terminating a flexible cable the outer _____ is scored using a _____ around the _____ of the cable taking care not to cut _____ the sheath.

● The sealing pot of a MIMS cable should be filled with sealing _____ from one side _____ to prevent _____ pockets being trapped inside.

● Where metallic armoured cables are terminated _____, continuity of the armour must be _____ and one method of doing this is by using an _____ locknut.

● The _____ and simplest method to terminate _____ metallic braiding is to use _____ sleeving.

● Fire resistant cables consist of an _____ sheath over a layer of _____ foil insulated _____ conductors and a _____ circuit protective conductor.

● Coaxial plugs and sockets are used with _____ cables where the _____ is connected to the _____ metal and the core to the _____ pin.

● The types of termination for fibre optic cables include _____ tip, _____ connector and _____ connector.

- Insulation _____ terminations are used extensively for _____ cables.

- Common terminations for fuse carriers and circuit breakers include _____ plate and screw _____ terminals.

- A crimp lug must be selected to _____ the conductor _____ and _____ in order for a good _____ to be made.

- Where printed _____ boards are made as a replaceable unit they often incorporate a _____ arrangement on the board such as an _____ connector.

- The cores of _____ cable will need to be _____ together and _____ in preparation for soldering.

- When soldering conductors, care must be taken to avoid damage to the _____, hardening of the _____ making it _____, and damage to _____.

LEARNING OBJECTIVES

On completion of this chapter you should be able to:

- Specify the health and safety requirements appropriate to terminating and connecting conductors, cables and flexible cords in electrical wiring systems and equipment, including:

  - Adherence to relevant statutory and non-statutory regulations

  - Selection and use of tools

  - Personal protective equipment (PPE)

  - Risk assessment

  - Reporting of unsafe situations.

Part 1 Health and safety requirements

Complying with the applicable legislation, standards and codes of practice when terminating and connecting conductors, cables and flexible cords in electrical wiring systems and equipment is essential.

The legislation which we must observe includes:

- The Health and Safety at Work (etc.) Act, and
- The Electricity at Work Regulations.

BS 7671, the Requirements for Electrical Installations, is published by the Institution of Engineering and Technology (IET) and is commonly known as the Wiring Regulations. Whilst these regulations are not statutory they are accepted as standard practice for electrical installation work. They may be referenced in a court of law should a prosecution be brought under the statutory regulations.

The British Standards Institution (BSI) produces other standards and codes of practice.

These statutory requirements, standards and codes of practice relate to the manufacture, installation and use of electrical equipment. Individuals should be properly qualified, trained and competent before working on electrical equipment.

A competent person is defined by the Construction (Health, Safety and Welfare) Regulations 1996 as:

> Any person who carries out an activity shall possess such training, technical knowledge or experience as may be appropriate, or be supervised by such a person.

Health and Safety at Work (etc.) Act 1974

The Health and Safety at Work (etc.) Act applies to everyone who is at work and it sets out what is required of both employers and employees. The aim of this Act is to improve or maintain the standards of health, safety and welfare of all those at work.

Figure 5.1 *Not good practice – the shutters are there for a reason*

A number of regulations and codes of practice have been introduced under the Health and Safety at Work (etc.) Act, including:

- Management of Health and Safety at Work Regulations
- The Electricity at Work Regulations
- Manual Handling Operations Regulations
- Control of Substances Hazardous to Health (COSHH) Regulations
- Workplace (Health and Safety and Welfare) Regulations
- Personal Protective Equipment at Work Regulations (PPE)
- Provision and Use of Work Equipment Regulations (PUWER)

- Work at Height Regulations
- Display Screen Equipment at Work Regulations
- Control of Asbestos at Work Regulations.

Any of these regulations which are applicable to either the termination and connection procedure, or to the environment in which it is undertaken, will need to be complied with when terminating and connecting conductors, cables and flexible cords in electrical wiring systems and equipment.

Remember

Watch out for new laws regarding health and safety and for amendments to the existing laws.

Tools and work equipment

Under the Health and Safety at Work (etc.) Act, employers are required to provide and maintain suitable safe tools and equipment for use by their employees. Employers will not generally provide the basic hand tools used by electricians as these are considered the employee's responsibility. However, any specialist tools such as crimpers, power drills and so on may be provided by the employer. Where necessary, training in the use of any such equipment must also be provided. Any other information or supervision that may be required is also the employer's responsibility.

Work equipment should be suitable for its intended use and for the conditions in which it is to be used. It should be maintained in a safe condition and inspected periodically to ensure that it remains safe for use. Records should be kept of the inspections carried out on the equipment.

The employer must also ensure that a safe method of working is in place and employees are aware of these procedures. Protective equipment, for example machinery guards, safety screens and protective clothing, must be provided where they are required.

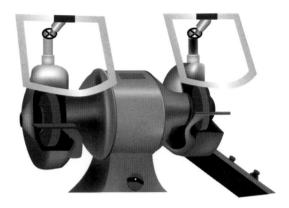

Figure 5.2 *Grinding machine*

An employee can be prosecuted for breaking the health and safety laws.

Employees are required by law to:

- Not interfere with tools, equipment, etc. provided for their health, safety and welfare
- Correctly use all work items provided in accordance with instructions and training given to them.

Your place of work could be an environment which has sources of hazard which need to be considered before starting work. For example, you may be asked to work in an area where others are drilling or chasing walls, using toxic substances and so on. Your work may involve terminating conductors in a confined space where access is limited. Other possible sources of hazard to be considered include the risk of being hit by falling objects and a lack of suitable lighting in the working area.

When work has to be done in an area that may be hazardous, consideration must be given as to whether the workplace can be made safer and whether there are some hazards that can be removed or reduced. Barriers may need to be erected, protective equipment worn or training undertaken.

Try this

List four statutory publications which relate specifically to the provision of work tools to employees:

1 _____

2 _____

3 _____

4 _____

In the Simple guide to the Provision and Use of Work Equipment Regulations 1998 there is an example list of equipment which is used by an employee at work. Make a note of the items listed and add any extra equipment that you may use when at work.

_____

_____

_____

_____

Part 2 Selection and use of tools

Appropriate tools should be selected for the task being undertaken and in this case we are concerned with the termination of cables and conductors. In Chapter 4 we considered the various methods of termination and so we shall take a look at the tools which may be required to carry out these terminations. Many of the tools come under the heading of general electricians' hand tools whilst others will be more specialist.

Connecting and terminating single and multi-core thermoplastic and thermosetting insulated cables will require a general tool kit including:

- Stripping knife (not a Stanley Knife type)
- Side cutters
- Electricians pliers
- Long nose pliers
- Strippers
- Terminal screwdriver (small)
- Terminal screwdriver (large).

Ripley Tools

Courtesy of Draper Tools Ltd

Figure 5.3 *Typical adjustable cable strippers*

Courtesy of Draper Tools Ltd

Figure 5.4 *Electricians' long nose pliers*

Many of these tools will be used when terminating other types of cables and we will not repeat them for each type in this chapter.

When it comes to terminating MIMS cables there are some additional hand tools to those listed above and a number of specialist tools which will be needed. These include:

- Junior hacksaw
- Measuring tape
- Small file
- Gland pliers
- Spanners
- MIMS stripping tool (rotary stripper)
- MIMS pot wrench
- MIMS crimping tool.

These specialist tools are available in various sizes to suit the cable being installed. The rotary stripper is adjustable to suit a number of sizes of cable, having different diameter holes for the different sizes.

Terminating armoured cables will not generally require specialized tools for glanding off the cable. There are some tools available to make termination of smaller cables easier to complete. The general requirements will include:

- Hacksaw
- Junior hacksaw
- Knife
- Spanners
- Gland pliers (two pairs) particularly for larger cable glands.

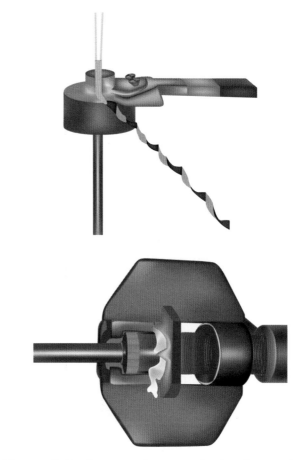

Figure 5.5 *Rotary stripper and crimping tool*

www.rel-Group.com

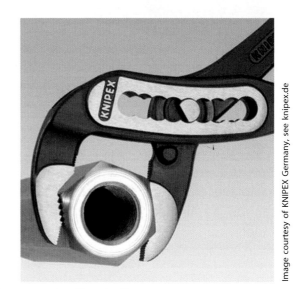

Image courtesy of KNIPEX Germany, see knipex.de

Figure 5.6 *Gland pliers*

For the smaller sizes of armoured cable a cable stripper may be used which minimizes the risk of damage when cutting through the armour.

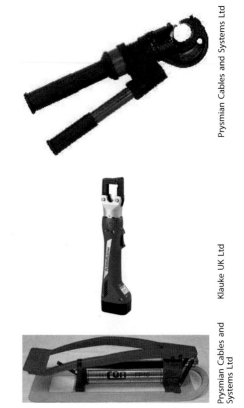

Prysmian Cables and Systems Ltd

Klauke UK Ltd

Prysmian Cables and Systems Ltd

Figure 5.9 *Hydraulic crimp tool*

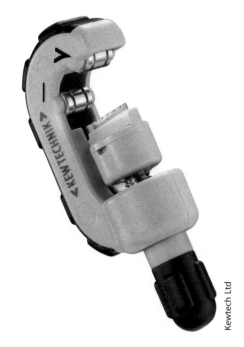

Kewtech Ltd

Figure 5.7 *SWA stripping tool*

Courtesy of Draper Tools Ltd

Figure 5.8 *Hand ratchet crimp tool*

For the larger sizes of armoured cable and termination to certain types of equipment we may need to use a crimp tool to terminate the conductors to the equipment. Smaller csa conductors can be terminated using a good ratchet crimp tool.

For larger csa conductors a hydraulic crimping tool may be required and these are available with dies to suit the conductor csa and shape. The correct dies must be used. Many companies hire these crimp tools as initial cost and the need for a variety makes this an expensive tool if it is only used occasionally.

There are three main types available:

a. Hand operated which can prove difficult if the termination is to be carried out in a confined space such as a distribution cabinet

b. Battery operated which are more manoeuvrable and do not require so much space to operate

Try this

You are to carry out an electrical installation which will include thermosetting single and multicore cables together with some small swa cables. Compile a list of the tools you will require to carry out the termination of these cables.

_____

_____

_____

_____

_____

_____

_____

c. Foot operated which have the advantage of needing only the crimp head attached to the hydraulic hose to be within the terminal location.

When terminating data cables there are some additional tools that will be required and these will depend on the type of cable being terminated. In general, there are basic tools for termination of cables together with some of the more specialist tools.

For data cables these may include:

● Cable stripper

Figure 5.10 _Wire stripper and cutter for single and ribbon cables_

● Ribbon cable cutter

Molex Incorporated

Figure 5.11 _Ribbon cable cutter_

● Flat ribbon cable crimping tool

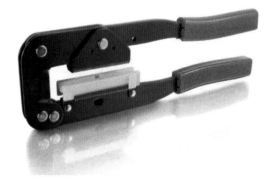

Figure 5.12 _Flat ribbon cable crimping tool_

- Insertion tool

Screw Fix

Figure 5.13 *Insertion and removal tool*

- Insulation displacement termination tool with blade
- Coaxial cable stripper

Labgear

Figure 5.14 *Coaxial cable stripping tool*

- Utility knife
- Soldering iron

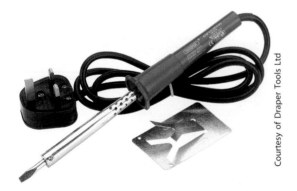

Courtesy of Draper Tools Ltd

Figure 5.15 *Soldering iron*

- Desoldering gun or braid
- Connectix tester

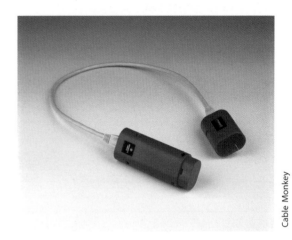

Cable Monkey

Figure 5.16 *Connectix tester*

And when terminating fibre optic cables:

- F-crimper
- F-connector
- Kevlar scissors

Figure 5.17 *Kevlar shears*

- Crimper
- Polishing puck
- Inspection microscope

Once the electrical installation work is complete it will need to be tested and this will require test instruments specifically for this task. These will include:

- Low reading ohmmeter
- Insulation resistance ohmmeter
- Earth fault loop impedance tester
- RCD tester
- Approved voltage indicator.

Note

The inspection and testing of the electrical installation prior to it being placed in service is covered in detail in the *Inspection Testing and Commissioning* study book in this series.

It is important to use the correct tools for the job you are undertaking and to use them as they are intended to be used. Misuse and mistreatment of tools is a major contributor to accidents in the workplace and one which can be easily avoided.

Tools should be checked regularly and any damaged tools must be repaired or replaced. Tools are expensive to replace and improper use can make them irreparable. Simple precautions can extend the life of your tools considerably and maintenance improves their operation and longevity.

A few simple rules:

- Only use tools for the purpose they are intended, for example using a screwdriver as a chisel will weaken the handle and damage the blade. When you next come to apply pressure to the screwdriver for that stubborn screw the handle may well split and the end of the blade could drive into your hand.
- Regular maintenance is a must: for example, a little light oil on the pivot hinges of

Cat No.45658 Fibre Optic Termination Kit, courtesy of Greenlee Textron Inc.

Figure 5.18 *Inspection microscope*

- Fibre buffer strippers

As well as the tools for termination we will also need some tools for the fixing and securing of the cables as this is often part of the final termination process. Typically we may need:

- Pin hammer
- Screwdrivers of various types and sizes
- Rubber mallet
- A power drill
- Masonry drill bits.

It is often necessary to put in the final cable fixings once the termination is made off as this allows some flexibility when terminating the cable. The final fixing is then necessary to ensure the cable is securely held in place and no stress is placed upon the termination.

sidecutters and pliers keeps them moving freely. This not only makes using them easier, it also means they are less prone to seizing up and to slipping during use.

- Make regular checks on the blades of screwdrivers to ensure the blades are square and sharp.

Figure 5.19 *Correct use of tools is important*

- Keep knife blades clean and honed to avoid slips and cuts. Always cut away from you and do not place any part of you in front of the blade.

- Make sure hacksaw blades are sharp and inserted the right way (teeth facing forwards). These need to be replaced fairly regularly and the correct type and cut of blade should be used for the job in hand.

Remember

Damaged tools must be repaired or replaced for your own and others' safety. Tools are expensive to replace. Take care of your own and your employer's tools, and keep them clean and in good condition. Tools must be kept in good condition and repaired or replaced when worn or damaged.

Task

Compile a list of the hand tools that you have in your tool box and then list here the additional tools you would need to complete the termination of a:

a MIMS cable.

_____

_____

_____

b Ribbon cable.

_____

_____

_____

c Resistor to a PCB.

_____

_____

_____

Part 3 Personal Protective Equipment

Personal Protective Equipment (PPE) is the final step in risk management. It is only used when there is no suitable alternative to reduce or remove the risks associated with a work activity. Wherever there are risks to health and safety that cannot be adequately controlled by any other means then PPE must be supplied and used.

Remember

You have a duty to ensure that you know, understand and use the equipment provided correctly. If the equipment is provided it must have been assessed as necessary. Never fail to use the PPE provided because the job will only take a few minutes or because it looks daft or unfashionable.

When we are terminating and connecting conductors, cables and flexible cords in electrical wiring systems and equipment there may be a need for PPE related to the activity. We shall consider some of the PPE equipment which may be necessary for the termination and connection activities. We will not consider the PPE equipment which may be necessary due to other site conditions or the activities of others.

For the termination and connection activities the following PPE should be considered.

Eye protection

Eye protection, by using safety spectacles or goggles, may be required in situations such as where there is a risk from flying debris, for

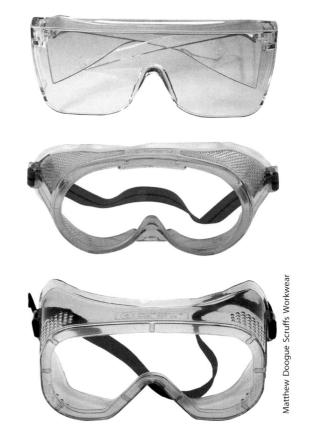

Figure 5.20 *Eye protection*

example bits of glass from fibre optic cables and particles of cut copper from copper cable.

Hard hats

Head protection, more commonly referred to as hard hats, may be required where there is a risk of injury from fixed protrusions in the work area, such as mounting brackets and pipe clamps whilst carrying out the termination of cables and conductors.

Matthew Doogue Scruffs Workwear

Figure 5.21 *Head protection*

Footwear

Protective footwear should be worn where there is a risk of injury from manual or mechanical handling, electrical work or any work carried out in hot or cold weather conditions. Many companies require their employees to wear safety boots or shoes on site at all times. When choosing footwear, consideration should be given to:

● Grip

● Resistance to water or hazardous substances

● Flexibility, and

● Comfort for the wearer.

Safety footwear is available in a variety of styles from heavy duty rigger boots to safety trainers.

Matthew Doogue Scruffs Workwear

Figure 5.22 *Safety trainers*

The type and style to be worn needs to be suitable for the environmental conditions on site.

Clothing

Work clothing needs to be appropriate for the activities being undertaken. In clean and dry indoor locations standard work trousers and tops are normally perfectly acceptable. In adverse conditions, such as when working outside, additional outdoor clothing may be required. Baggy sleeves and loose tops are not considered appropriate when carrying out the termination of cables and conductors.

Kneeling pads

The termination of cables and conductors often involves kneeling for some time and to protect your knees from long-term damage knee pads should be worn. These may be separate knee pads worn over the trousers or overalls. Many manufacturers of work trousers include pockets in the trousers to take the knee pads which means there are no additional straps around the legs and are often found to be the most comfortable way of providing knee protection.

Matthew Doogue Scruffs Workwear

Screw Fix

Courtesy of Draper Tools Ltd

Courtesy of Draper Tools Ltd.

Figure 5.23 *Knee pads and work trousers with knee pad pockets*

Figure 5.24 *Hand protection*

Gloves

Leather gloves can safeguard against cuts resulting from manual handling of heavy and sharp objects. Gloves may be required when working with armoured cables and the associated materials where there is a risk of injury to the hands from sharp edges, tape and wire armour.

When working with chemicals as part of the termination process, such as lubricants, corrosion inhibitors and jointing compounds, gloves should be used to avoid contact with the skin and the long-term effects of such exposure.

When working in a location where there is a risk of an electric shock, electric arc or burns because equipment in the close vicinity cannot be made dead, then wearing suitable gloves may provide the protection required.

Try this

You are going to terminate cables on a new leisure centre complex. List the potential risks that you may encounter when carrying out the actual terminations (ignoring any environmental issues) and any appropriate PPE equipment required for the following situations:

a A number of swa cables with small csa copper conductors in the bases of the car park lighting columns.

b Crimped large csa aluminium conductor lugs in the main distribution board.

c Flat twin and cpc cables at the plug in ceiling roses for the office lighting.

a Risks: _____

PPE: _____

b Risks: _____

PPE: _____

c Risks: _____

PPE: _____

Part 4 Risk assessment

A risk assessment needs to be carried out when terminating and connecting conductors and cables to determine whether any of the work activities are likely to cause harm.

Note

Risk assessment is covered in more detail in the Organizing and Managing the Work Environment study book in this series.

The objective of the risk assessment process is to identify and reduce the risks where it is 'reasonably practical' to do so. Reasonably practicable is a term used in legislation and it means that an employer must balance the cost of the actions taken to reduce the risk against the degree of risk which exists. This will include the trouble, effort and time taken and not just the expense.

Figure 5.25 *Risk assessments*

The risk assessment should consider everyone who may be affected by the activity, not just those carrying out the work. This will include employees, other contractors, members of the public and visitors to the site.

The way in which a risk assessment is carried out will vary from one organization to another with each having a slightly different approach. What is important is that a risk assessment is carried out and all the foreseeable risks are considered.

To make a risk assessment it is important to understand the difference between a risk and a hazard. In general terms we can consider them to be:

● *Hazard*: A hazard is anything with the potential to cause harm. Electricity, noise and hazardous substances are typical examples of hazards.
● *Risk*: A risk is the likelihood of a hazard causing injury, damage or loss and how severe the outcome may be.

Most companies will have carried out risk assessments for the everyday activities of their staff and this would include the termination of cables and conductors. The company safety manual will include the risk assessment and identify the procedures and any associated PPE for these activities. You will need to follow the company's procedures when carrying out these tasks. It is also your responsibility to carry out a risk assessment for each activity to ensure that the procedures and PPE are appropriate.

Your risk assessment should include:

● Identify the hazards associated with the work activities
● Identify who could be harmed by those hazards
● Identify how you manage the risks.

Each activity should be considered critically, including how it is carried out. Reference to any existing guidance and information may help in determining the risk.

Remember

We all make risk assessments every day. Before we cross the road we make a risk assessment for the presence, speed and distance of any vehicles and determine whether we are able to cross safely. Risk assessments at work are an extension of this process.

To decide on the most appropriate measures to deal with the risk we would consider the following points:

- Can the risk be eliminated altogether?
- Can the risk be contained by additional procedures or equipment?
- Can the work be adapted or arranged to suit a particular situation?
- Is there a technological or engineering alternative to carrying out the work?

The majority of these issues will have been considered in the generic risk assessment and the need for equipment and special tools will be identified. The additional provision of PPE is a last resort when there are no other control measures which can be used.

When you receive a risk assessment the information may be in a format similar to that shown in Figure 5.26 and you will need to ensure the requirements are complied with.

Suppose you are terminating the conductors for a new power circuit into an existing distribution board in the switchroom of an office building. Then typical risk assessment information would be:

Work activity: Termination of cables

Hazards:

1 Contact with live parts, short circuit
2 Flying debris
3 Cuts.

Likely harm: Electric shock to operative or other people, electrical burns and injury from flying debris and sharp materials

Risk rating: High (if uncontrolled), low (if controlled)

Control measures:

- Isolation of exposed live parts
- Use of barriers to cordon off the area to other personnel
- Provision of eye protection and work gloves
- Working to be in accordance with requirements of DM Associates Health and Safety Guidance.

Additional requirements:

Only DM Associates quality assured and calibrated instruments to be used.

DM Associates Ltd Risk Assessment						No.	
Assessed by		Date		Authorised by		Date	
Work activity:							
Task	Hazard	Likely harm	Risk rating	Control measures required	Additional requirements	Further information	

Figure 5.26 *A typical risk assessment form*

Further information:

Further guidance can be found in:

- HSE Guidance GS 38
- IET Guidance Note 3 Inspection and testing
- Memorandum of Guidance on the Electricity at Work Regulations HS (R) 25
- Reference to client's safety briefing.

The risk assessment has set out the nature of the task, the likely hazards and the possible harm that could result. The level of risk is given in the risk rating and without the control measures, the risk is high. If the control measures are implemented the risk is reduced to low. The control measures are then given and these may identify the need for a level of competence and use of correct procedures and equipment. This may require a certain level of training and demonstrable competence.

Try this

A lighting circuit plug in a ceiling rose for fluorescent luminaires is to be terminated in a suspended ceiling at high level in the vacant floor of an office block. The termination is to be installed before a suspended ceiling is erected.

List all the work activities and the associated hazards which would need to be addressed in a risk assessment before this work can be carried out.

- _____
- _____
- _____
- _____
- _____

Part 5 Unsafe situations

Every employee is required to take reasonable care of their own health and safety and that of others.

Employees are therefore required to look out for hazards in the workplace and report those that they find. If a hazard is ignored this may cause an injury to you or others.

Hazards will include such things as deficiencies in tools and equipment, for example broken ladder rungs, insecure structures and falling objects.

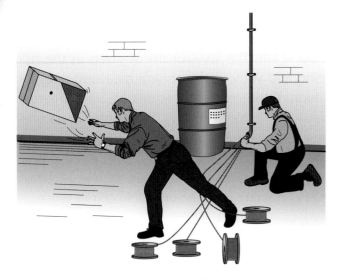

Figure 5.27 *Work safely and sensibly*

Deficiencies in equipment should be reported as soon as they are noticed. For example, a broken ladder rung or a damaged hose on a hydraulic crimp tool are hazards that should be reported before they cause injury to a user.

Figure 5.28 *Always report potential hazards before they cause injury!*

Create a safe working environment so when, for example, terminating cables into a sump pump

the area around our workspace should be safe and secure.

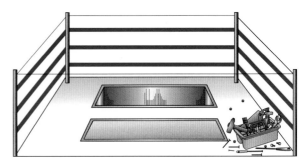

Figure 5.29 *Securing the area around the sump pump termination pit*

Accidents should always be reported whether they result in injury or not as it may prevent another more serious accident from the same cause. Near misses should also be reported and recorded so that the incident can be reviewed and any further precautions implemented.

Any reports should be made to an appropriate responsible person. This may be your employer, supervisor or the health and safety representative.

Where the situation is within your control and the action to correct it is straightforward, the situation may be readily corrected. However, a report should still be made to identify the incident, the cause and the action taken.

Where the action is outside your control the report must be made to the person responsible without delay and a temporary measure can be taken whilst this is carried out. Simply getting someone to stand by an open access pit would enable others to be warned and avoid an accident.

DM Associates Ltd		
Unsafe Occurrence Report Form		
This form must be completed in the event of any unsafe occurrence at (Site Location).		
Date and time of observation	*Date:*	*Time:*
Location of the incident		
Brief description of the equipment and or circumstances:		
Details of any action taken		
Reported by:	*Name:*	*Job title:*
Received by	*Name:*	*Job title:*
Action Taken		
Date completed	*Date:*	*Time:*
Authorised by	*Name:*	*Job title:*

Figure 5.30 *Unsafe situation report form*

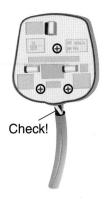

Check!

Figure 5.31 *Visually check cable and plugs*

Equipment hazards

When portable electrical equipment is used in the workplace, or on construction sites, accidents can often be prevented by following a few simple rules. A visual check on cables and plugs, which are particularly liable to damage, can prevent a serious accident.

Equipment may develop faults which do not affect its operation, yet the equipment may present a potential hazard. Regularly inspecting the equipment and identifying any necessary maintenance and repair helps to ensure accidents are prevented.

Task

Whilst at work you inspect the tools provided for the work you are expected to do that day and you notice a fault that could cause a potential hazard.

Identify a fault on an item of equipment and complete the form in Figure 5.30.

Try this: Crossword

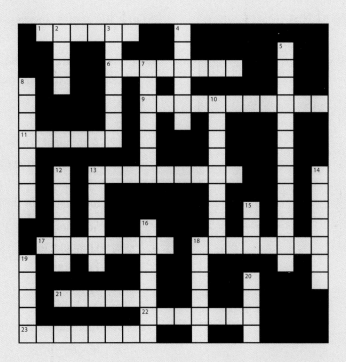

Across

1   This is a bit of a pull but a pot version is used with MIMS terminations (6)

6   Where the conductor finishes (8)

9   When you place the conductor in 6 across you are doing this (11)

11 Keep the cuts at bay and protect your digits (6)

13 We must comply with requirements which are this (9)

17 Boots and trainers are both this (8)

18 These protect your leg joints (4,4)

21 Strikes a blow when clipping cables (6)

22 If not protected cable may be ... (7)

23 Use these to tighten nuts, bolts and glands (8)

Down

2   The chance of being affected by a 16 down (4)

3   Do electricians use these on the side to cut cable? (7)

4   Doubling over solid conductors is done with these (6)

5   Not grown up but used to cut armoured cables (6,7)

7   This type of crimp tool ensures the correct pressure is applied (7)

8   A pointy version of 4 down (4,4)

10 The type of tool used for IDC (9)

12 Worn in the hair or to transfer data (6)

13 A lack of danger (6)

14 Keep your vision protected with these (7)

15 The last line of defence in managing risk (3)

16 These may endanger us at work (7)

18 This is used in fibre optic cable and trimmed with shears (6)

19 Another 17 across (5)

20 One of these is either secret or a ... of practice (4)

Congratulations, you have now completed Chapter Five, the last chapter in this study book. Complete the self assessment questions before continuing to the end test.

SELF ASSESSMENT

Circle the correct answers.

1   To terminate a multicore cable at a three plate ceiling rose the most appropriate tool for the task is:

   a.   An IDC tool

   b.   A crimping tool

   c.   A terminal screw driver

   d.   A medium screwdriver

2   The tool used to fit the pot on to a MIMS cable is a:

   a.   Stripping tool

   b.   Crimping tool

   c.   Pot wrench

   d.   Spanner

3   The tool shown in Figure 1 is a:

   a.   Ribbon cutter

   b.   Universal stripper

   c.   Ratchet hand crimper

   d.   Multi-conductor IDC tool

4   The most appropriate PPE requirements whilst terminating a number of cables at low level will include:

   a.   Safety boots

   b.   Knee pads

   c.   Hard hat

   d.   Gloves

5   An unsafe situation which can be readily corrected must be:

   a.   Reported

   b.   Remedied without reporting

   c.   Barriered off and left to others

   d.   Not reported if this delays the work

End test

Tick the correct answer

1. **To confirm that a three plate ceiling rose is safely isolated, a number of voltage tests will need to be carried out at the ceiling rose. Which of the following tests is not required?**

   ☐ a. lines to neutral

   ☐ b. line 1 to line 2

   ☐ c. lines to earth

   ☐ d. neutral to earth

2. **An approved voltage indictor (AVI) has a voltage range 6V to 1000V ac. On which of the following ac systems would it be very dangerous to use this AVI?**

   ☐ a. HV

   ☐ b. LV

   ☐ c. ELV

   ☐ d. SELV

3. **When confirming safe isolation of a single phase circuit the total number of tests to be made is:**

   ☐ a. 3

   ☐ b. 4

   ☐ c. 7

   ☐ d. 10

4. **Once safe isolation has been confirmed the test equipment must be confirmed to be:**

   ☐ a. in calibration

   ☐ b. compliant with GS 38

   ☐ c. working using a proving unit

   ☐ d. suitable for the expected voltage

5. **Where work has to be carried out on live equipment the Electricity at Work Regulations requires that:**

   ☐ a. the DNO is notified of the work

   ☐ b. the client is notified of the work

   ☐ c. work is done as quickly as possible

   ☐ d. suitable safe procedures are followed

6. **The numeral 6 in BS 7671 Regulation 526.1 indicates that the regulation can be found in:**

   ☐ a. Part 6

   ☐ b. Section 6

   ☐ c. Chapter 6

   ☐ d. Regulation 6

7. **Which of the following statements does not apply to an Approved Code of Practice?**

☐ a. they are non-statutory

☐ b. they provide guidance

☐ c. they may be cited in a court of law

☐ d. they may not be cited in a court of law

8. **A Variation Order is used to notify contractors of:**

☐ a. changes to the technical specification

☐ b. additional installation work

☐ c. site working hours

☐ d. salary increases

9. **Which of the following is not a requirement for all electrical terminations?**

☐ a. mechanically sound

☐ b. electrically sound

☐ c. suitably fixed

☐ d. accessible

10. **A junction box which does need to be accessible must have the manufacturer's marking:**

☐ a. NF

☐ b. FN

☐ c. MF

☐ d. FM

11. **Corrosion at terminations may occur as a result of:**

☐ a. dissimilar metals

☐ b. incorrect torque

☐ c. overtightening

☐ d. similar metals

12. **Every cable entry into an enclosure must have:**

☐ a. a stuffing gland

☐ b. protection to IP6X

☐ c. colour identification

☐ d. mechanical protection

13. **Which of the following is not a consideration when selecting a termination for a conductor?**

☐ a. the number of wires which make up the conductor

☐ b. the number of conductors to be connected together

☐ c. will vibration be present in normal service?

☐ d. is the conductor accessible in normal use?

14. **A thin blade trimming knife should not be used to strip the sheath from a flat profile multicore cable because it:**

☐ a. can easily damage the conductor insulation

☐ b. is not sharp enough to cut the sheath

☐ c. can only be used for circular cables

☐ d. does not glide easily when cutting

15. **The outer sheath of a MIMS cable is best removed using:**

☐ a. an automatic cable stripper

☐ b. a pair of sidecutters

☐ c. a stripping knife

☐ d. a rotary stripper

16. A steel wire armoured cable must be terminated using a gland which:

☐ a. provides an anchor for the protective shroud

☐ b. allows the cable to flex with any vibration

☐ c. securely holds all the wires of the armour

☐ d. stops the outer sheath from fraying

17. When terminating an aluminium armoured cable it is important that the armour is:

☐ a. visible before it enters the shroud

☐ b. securely and reliably earthed

☐ c. soldered to the gland body

☐ d. secured in a brass gland

18. When terminating fibre optic cables care must be taken to ensure the glass offcuts are:

☐ a. thrown in the general waste

☐ b. wrapped in paper for landfill

☐ c. put into a sharps container

☐ d. left to be swept up later

19. The terminal shown in Figure 1 is a:

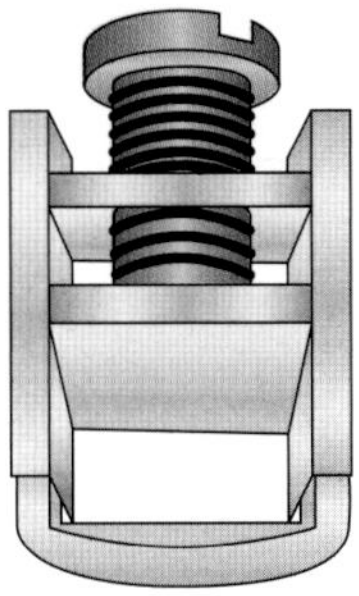

Figure 1

☐ a. screw pinch

☐ b. pressure plate

☐ c. wraparound

☐ d. tunnel

20. A crimp lug must be selected to suit the conductor shape and:

☐ a. temperature

☐ b. diameter

☐ c. length

☐ d. csa

21. The termination shown in Figure 2 may be used to terminate an aluminium conductor to a:

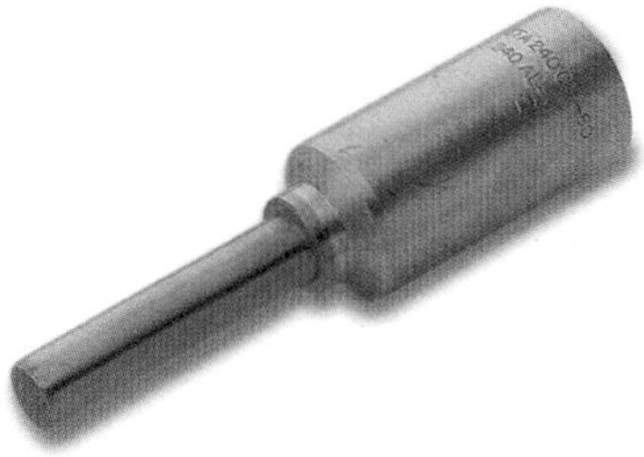

Figure 2

☐ a. bus bar

☐ b. brass terminal

☐ c. bolted connection

☐ d. wraparound terminal

22. Connections to small electronic components are generally made with a soldering iron and:

☐ a. flux

☐ b. tinning

☐ c. plain solder

☐ d. multicore solder

23. Before soldering a stranded conductor to a tag connection the strands should be twisted together and:

☐ a. shaped to the terminal

☐ b. dipped in flux

☐ c. squeezed flat

☐ d. tinned

24. Tools used for the termination of flat twin and cpc cables must be:

☐ a. maintained in good condition

☐ b. provided by the company

☐ c. approved by the HSE

☐ d. not insulated

25. The tool used to fit the seal into the pot of an MIMS cable is a:

☐ a. stripping tool

☐ b. crimping tool

☐ c. pot wrench

☐ d. spanner

26. The tool shown in Figure 3 is:

☐ a. a pair of universal pliers

☐ b. an adjustable spanner

☐ c. a pair of gland pliers

☐ d. a pair of pliers

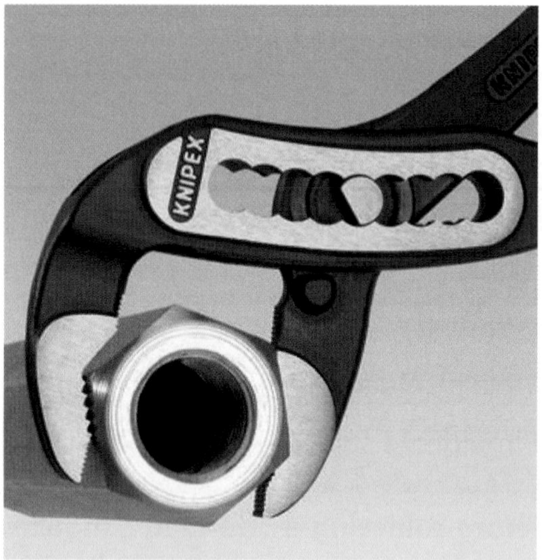

Figure 3

27. Which of the following tools is not specifically related to the termination of fibre optic cables?

☐ a. inspection microscope

☐ b. polishing puck

☐ c. Kevlar scissors

☐ d. crimping tool

28. When carrying out soldering operations, PPE should be used to prevent contact with:

☐ a. aluminium conductors

☐ b. hot soldering tip

☐ c. copper sulphite

☐ d. soldering flux

29. Before carrying out the termination of a number of large cables at a distribution board which is energized and in use, the electrician should make:

☐ a. a risk assessment

☐ b. application to the DNO

☐ c. the office aware of the work

☐ d. the HSE aware of the work

30. Unsafe situations arising from faulty equipment can best be prevented by:

☐ a. repairing equipment when it becomes faulty

☐ b. regular inspection of equipment

☐ c. replacing equipment annually

☐ d. the use of PPE

Answer section

Chapter 1

Task Page 4

2 a. The cardiac muscles of the ventricals
    b. Flat line with no heart activity at all:
      http://en.wikipedia.org/wiki/Asystole
    c. Cardiac death

Task Page 7

2 a. inadequate space to work safely
    b. an insecure footing
    c. insufficient light
    d. potentially flammable gases or vapours
    e. explosive or conductive dusts

Task Page 12

a. cracked meter cases
b. damaged insulation (abrasion, cuts or perishing of flexible insulation)
c. loose terminals

SELF ASSESSMENT Page 14

1 a. only statement (i) is correct
2 c. GS 38
3 b. 0.05A
4 d. 10
5 c. Two padlocks, multi-lock hasp and two warning notices

Chapter 2

Recap Page 16

- 50V 120V dc Health Safety GS 38
- more 4mm exposed 2mm finger contact live
- ensuring switched secured inadvertently
- safely installations serious shock others
- burns entry exit contact within path
- persons equipment services
- line neutral
- line earth
- Neutral earth
- line

voltage any

Task Page 20

526.2

Task Page 23

1 BS EN 61558-2-5
2 BS 1363
3 BS 67
4 BS 1362
5 BS EN 60238

Try this Page 28

ORDER

Wordsearch Solution

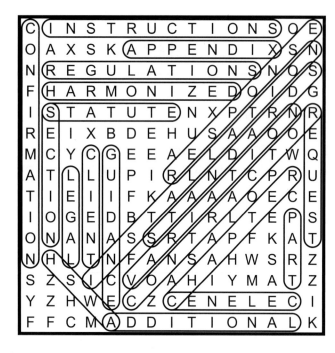

SELF ASSESSMENT Page 29

1   c. Electricity at Work Regulations
2   a. Part 5
3   b. Is relevant to a matter of prosecution
4   d. Variation order
5   c. The name of the wholesaler

Chapter 3

Recap Page 30

●   enforceable   law   cited   court
●   all   statutory

●   Part   Chapter   Section
●   non-statutory   selection   erection installations
●   instructions   connection   torque
●   Variation   changes   specification
●   estimate   time   materials   quotation programme
●   instructions   latest   remain
●   variation   confirm   accepted

Try this Page 37

1   a. *Any from pvc grommet, stuffing gland, male bush and coupler*
    b. *Brass bush*
    c. *No additional protection as it is a plastic enclosure*
    d. *Suitable SWA gland*

2   a. *IP4X or IPXXD*
    b. *IP2X or IPXXB*
    c. *IPXXB*

Try this Page 44

1   *Dust and corrosion*
2   *Water, heat and chemicals*
3   *Water, salt, sand and animal waste*

SELF ASSESSMENT Page 45

1   a. Accessible for maintenance
2   c. IPXXD
3   a. There must be protection against mechanical damage
4   c. The position of the termination
5   a. Prevent corrosion

Progress Check Page 46

1   c. obtain permission to isolate
2   c. part of the installation will not function normally
3   b. a shock risk to you and others
4   c. before and after use
5   d. all live conductors and all live conductors and earth
6   d. their own unique padlock and key separate from all others
7   d. the supply can be isolated
8   c. Chapter 2
9   b. non-statutory but may be cited in a court of law
10 a. variation order
11 b. name of the wholesaler
12 d. variation orders
13 c. both statement i) and statement ii) are correct
14 d. a crimped joint in a cable
15 b. IP2X
16 c. held by a cord grip to prevent strain on the terminals
17 a. electrolytic action
18 d. identification of conductors
19 c. shape
20 b. vibration

Chapter 4

Recap Page 49

- appreciable   strain   termination
- MF   manufacturer's   accessible
- provides   provide   protection
- enclosure   cables   conductors
- barriers   solid   liquids
- insulated   accessory   luminaire   appropriate

- conductor   shape   wires   number   connected
- Dissimilar   corrosion   electrolytic   aluminium
- stress   clamps   glands   containment
- flexible   system   anti-vibration   cable

Try this Page 54

1   a. Damage the insulation
    b. Damage the conductor
    c. Cut the sheath back too far
    d. Cut the insulation too deep so that you damage the conductor

Try this Page 58

Because the insulation is hygroscopic and will absorb moisture if not sealed correctly.

Try this Page 64

Using either a banjo beneath the locknut and a protective conductor to the earthing terminal, or using an earthing locknut and a protective conductor to the earthing terminal.

Try this Page 71

1   Fibre optic connectors
2   Carbide scribe
3   Kevlar shears
4   Lint free cloth
5   Pure alcohol solution for surface cleaning
6   Polishing surface (usually a glass plate and a rubber mat)
7   Polishing disc
8   Polishing papers (1, 3 and 5 microns)
9   Syringe for the epoxy adhesive

10   Epoxy adhesive
11   Adhesive activator
12   Fibre optic crimp tool
13   Fibre optic stripper
14   Fibre optic microscope

Try this Page 83

1   a.   Cleanliness
        Correct heat
        Correct flux.
   b.   Any three from:
        Dry joints
        High resistance joints
        Excess solder
        Damage to insulation
        Damage to the PCB
        Hardening of the wire making it brittle
        Damage to components

2   Use
     Desoldering braid
     A desoldering tool

Try this: Crossword Page 87

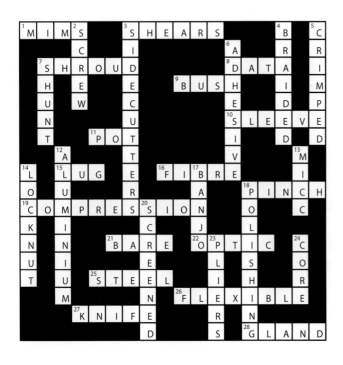

SELF ASSESSMENT Page 88

1   a. Scored with a sharp knife where the sheath is to be removed
2   c. Prevent the ingress of moisture
3   b. Halfway
4   d. Aluminium
5   a. An insertion tool
6   a. Screw pinch
7   d. Shape
8   b. Finishes at the terminal edge
9   b. Edge connectors
10   c. Heat shunt

Chapter 5

Recap Page 90

●   insulation   conductor
●   sheath   knife   circumference   right through
●   compound   only   air
●   earth   maintained   earthing
●   neatest   non   heat shrink
●   outer   aluminium   live   bare
●   screened   screening   outside   centre
●   straight   square   little
●   displacement   ribbon
●   pressure   pinch
●   suit   csa   shape   connection
●   circuit   plug-in   edge
●   stranded   twisted   tinned
●   insulation   wire   brittle   components

Try this Page 94

Any four from:

●   The Electricity at Work Regulations
●   Manual Handling Operations Regulations

- Workplace (Health and Safety and Welfare) Regulations
- Personal Protective Equipment at Work Regulations (PPE)
- Provision and Use of Work Equipment Regulations (PUWER)
- Display Screen Equipment at Work Regulations

Generally, any equipment which is used by an employee at work (PUWER) is covered, for example hammers, knives, ladders, drilling machines, power presses, circular saws, photocopiers, lifting equipment (including lifts), dumper trucks and motor vehicles.

Try this Page 97

Knife, pliers, side cutters, cable strippers, screwdrivers (terminal, medium, large and pozidrive), junior hacksaw, spanners and gland pliers. May also include SWA stripper, crimp tools and long nose pliers.

Try this Page 104

a. Sharp metal edges, flying debris from cutting and kneeling
   Work gloves, safety glasses and knee pads
b. Lubricant/corrosion inhibitors and kneeling
   Work gloves and knee pads
c. Flying debris from cutting above head height
   Safety glasses

Try this Page 107

- Working at height: use of access equipment, use of power tools
- Installing fluorescent lighting outlet: working above head height causing strain on muscles and back

- Injury from falling
- Eye injury
- Inhalation of masonry dust

Try this: Crossword Page 111

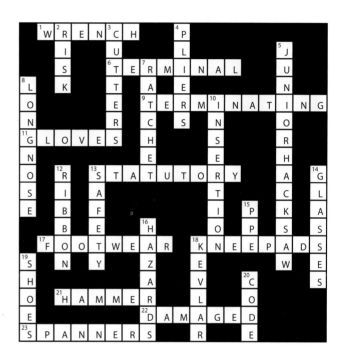

SELF ASSESSMENT Page 112

1 c. A terminal screwdriver
2 c. Pot wrench
3 c. Ratchet hand crimper
4 b. Knee pads
5 a. Reported

End test

1 b. line 1 to line 2
2 a. HV
3 a. 3
4 c. working using a proving unit
5 d. suitable safe procedures are followed
6 b. Section 6
7 d. they may not be cited in a court of law

8   a. changes to the technical specification
9   d. accessible
10 c. MF
11 a. dissimilar metals
12 d. mechanical protection
13 d. is the conductor accessible in normal use?
14 a. can easily damage the conductor insulation
15 d. a rotary stripper
16 c. securely holds all the wires of the armour
17 b. securely and reliably earthed
18 c. put into a sharps container

19 b. pressure plate
20 d. csa
21 b. brass terminal
22 d. multicore solder
23 d. tinned
24 a. maintained in good condition
25 b. crimping tool
26 c. a pair of gland pliers
27 d. crimping tool
28 d. soldering flux
29 a. a risk assessment
30 b. regular inspection of equipment

Glossary

ATA Aluminium tape armoured
AVI Approved voltage indicator
AWA Aluminium wire armoured

CENELEC European electrical standards body
Coaxial A coaxial cable has inner and outer cylindrical conductors
COSHH Control of Substances Hazardous to Health Regulations

DIL Dual-in-line

ELV Extra low voltage
EWR The Electricity at Work Regulations

FP and PX cables Fire resistant cables

Hazard A hazard is anything with the potential to cause harm
HSWA The Health and Safety at Work (etc.) Act

ID Insulation displacement
IDC Insulation displacement connection
IET Institution of Engineering and Technology
IP Ingress protection

Kevlar A strong synthetic fibre (Poly-paraphenylene terephthalamide) widely used as a protective outer sheath for optical fibre cable

LC Little connector
LED Light emitting diode

MF Maintenance free
MICC Mineral insulated copper clad cable
MIMS Mineral insulated metal sheathed

PCB Printed circuit board
PILC Paper insulated, lead sheath cable
PILCSTA PILC Steel tape armoured
PILCSWA PILC Steel wire armoured
PPE Personal Protective Equipment
PUWER Provision and Use of Work Equipment Regulations
pvc Poly vinyl chloride

Risk A risk is the likelihood of a hazard causing injury, damage or loss and how severe the outcome may be
RoHS Restriction of Hazardous Substances

SC Square connector
ST connectors Straight tip connectors
SWA Steel wire armoured

Ventricular fibrillation Where the heart rhythm is disrupted and results in irregular fluttering rather than the beat required to circulate blood around the body

WEEE Waste Electrical and Electronic Equipment Directive

XLPE Cross linked polyethylene

Index: Termination and Connection of Conductors